THE BOOK OF GEET

BOOK ONE
BASIC SCIENCE

THE BOOK OF GEET

BOOK ONE
BASIC SCIENCE

PAUL PANTONE

ISBN-10:0996688307

ISBN-13:978-0-9966883-0-7

Cover Design: Di Freeze

Cover Image: Sarawut Padungkwan

Contents

Foreword

More years ago than I care to think about, I was pointed toward Paul Pantone by an unremembered third party, describing him as a man of vision who had working inventions that would have a significant impact upon the world. Having never met the man, I called the provided phone number and we talked about his GEET system. I was hooked.

I asked how I could learn this amazing technology and he told me about his school. Putting food on the table and a shelter over my family's head took precedent so I could not absorb the cost of tuition. I didn't want this to slow me down as for years I have been a voracious consumer of all things 'alternate energy' so I asked him "What do you need?" He was having a time trying to communicate with his students as they could not take notes fast enough, nor make sketches of the concepts of GEET so I offered my services, drawings for lessons. It was agreed and we never looked back.

We seemed to be able to communicate well. He would put out a concept and I would do my best to bring it to life with visuals. For the most part, after some minor corrections every now and again, we put together enough visual information at the front of the classroom as well as in his student's hands, which improved his lesson retentions significantly. Instead of scrambling to take notes and draw diagrams, his students could focus on the instruction and make some personal notes on their copies of the drawings. After I provided the necessary drawings I asked him "So when do I get to take the class?" Paul replied, "Why? You already know more about GEET than any of my students who have come through this school."

For my part, I am happy to say it has been a privilege to work with Paul, because he can make my brain explode with ideas and practical applications. The hardest part about working with Paul and having him as a friend has been watching him go through the struggles, trials and tortures dealt by the very people we look to for leadership. Naysayers abound and professional naysayers seem to live to make a name for themselves instead of examining all the facts and printing real truth.

My son John joined me in exploration of GEET making know to his high school chemistry class the abilities of a GEET modified engine. He wanted to do a demonstration for his class, but had to spend quite some time after school with his chemistry teacher and the vice principal of the school going over the technology. Skeptical, they agreed and on a nice sunny day, on the lawn of the high school his chemistry class gathered round where we had set up a bare 3½ HP GEET modified engine. I remember his chemistry teacher gave a halfhearted kick at the engine head and said, "You probably have a fuel tank hidden inside here." I told the class this unit really likes to run on Mountain Dew. No one had any but there was some Pepsi so I asked for a donation, which the young lady provided. We poured it into the bubbler, sealed it up and pulled the rope. The little engine ran beautifully. I had some of the students smell the exhaust and feel the engine exhaust muffler. This is not something you can do with a conventional fuel engine as you will suffer from personal pollution as well as burn yourself on the muffler. Smiling, my son later told me his chemistry teacher had become a little less skeptical.

So now we come to The Book of GEET and my sincere hope, as you read through these varied and semi-technical writings, looking over drawings I have been doing over the years, that your mind, too, would be stimulated about possibilities and your eyes open to forces in this world who stand in fear of change, not wanting these possibilities to come to light.

Read, learn, take action and explore. Your world will never be the same.

Clarence Close

Introduction

The book of GEET is not just another science book. This book does not promote sales or investments. It delves into many aspects of freeing the world from pollution and energy control. I stopped counting the number of discoveries attributed to GEET in 2003. Back then, there were over 100 discoveries. I thought it was time to validate all I could for students around the world and people like me who are looking for answers but could not find them in the conventional books available through a library or university. If you have sought the answers to clean air and water with an unlimited supply of energy and could not find them, I am hoping you will find exactly that within the contents of this book.

Some of the discoveries made occurred over a hundred years ago. Where did the books go that had this knowledge? Did they simply vanish? Were they destroyed? We may never know the truth, but it is not relevant today. What is important is we cannot forget this knowledge. I hope others can expound upon it. If I can pioneer the road to clean fresh air and clean water by showing you how to reduce emissions from every single piece of equipment on the planet that uses fuel, even if only by 10 percent, this would be a major accomplishment.

Some of the things you will learn in this text will show you that 100 percent reduction in pollution is possible, and it has been accomplished. However, to be politically correct, in some cases, we will only be able to show improved efficiencies by 10, 20, or 30 percent. Governments around the world receiving political contributions from oil companies and major corporations cannot afford to turn their backs on those lucrative paychecks. Success and greed do not happen at the same time in this political environment. We must also keep in mind many investors are just Mom and Pop type of people.

Many inventors before me have faced the same turmoil, rejection, and persecution as well as prosecution for trying to help mankind. This inventor, for one, will get the knowledge out to the world in whatever way I can. Before going into how GEET works and all of the discoveries I have made, I would like to thank each and every person who has helped me along the way. Some of you showed love and respect, while others expressed hatred for the technology advancements I brought forward. I am still smiling.

Some reporters have done a good job in reporting the facts. Others want to make a name for themselves and print any story to help them accomplish their own recognition. Many of these writers and reporters covered products by other which simply do not work. They do not realize the damage they cause when they promote something which does not function properly. Their reporting hurts real inventors who have things that do work. One reporter went as far as to say that GEET works, but he did not see any way anyone would ever want to work with the inventor. Please report the facts about the inventions. The personal feelings of a reporter do not need to be a factor in the pursuit of science.

Other reporters have made up stories to bolster their own egos. Above all, the worst reporters I have met refused to cover the story if it did not help their church or religious beliefs. I will respect all people if they show me they are deserving of such respect. If your teaching says alcohol and tobacco and other things are bad, why would you sell those products at your store? There are only two stores I found in the United States, so far, that refused to sell tobacco and beer because it was against their religion. They deserve my respect and yours too. To do the job to help all of humanity, I need to be able to help all people who have, or do not have, a religion or faith.

The greatest pleasures I have are quite simple. One such pleasure was standing before a class of about 35 students. As I looked around, I could see several who did not speak English. Within the same group were Hindus, Buddhists, Christians, Amish, and atheists. During the entire week of class, they all got along like brothers. By the third day of class, everyone in the group had assembled a GEET demo engine, and every engine worked. It was at that time I realized language would not stop me from teaching.

Many times, I donated time to help people along the way. During the last 64 years of my life, I held many signs for many groups picketing pollution problems and other social issues I thought were very real. It is heartbreaking to find out that so many of these organizations only do it for the donations they receive and really could care less about the outcome. They put on a good show at our expense.

When I contacted many of these organizations in 2004, most did not have the decency to respond. A few sent back envelopes saying, "If you want to make a donation, use our envelope, send us your credit card information, and tell us whether it's weekly, monthly or yearly." What they were responding to was a video and over fifty pages of documentation to prove this new technology could reduce pollution to zero. Not one environmental foundation, anywhere in the world, was willing to spend a dime to provide to the world an actual proven solution.

Glossary of Terms

1) Auric: A created field around the rod resembling an aura around a human or planet tissue.

2) Biblical North: straight up from anywhere on the planet; biblical south is earth center.

3) Auric Field: The auric field exists in different layers. Sometimes referred to as harmonics because of the color fields they emit.

4)Auric Effect: There are many discoveries that have to be made of the possibilities of the auric effect, but mostly within this book it is one of the eight gravitational fields which create plasma.

5) North: to or towards the north

6) Magnetic North: The direction toward the magnetic North Pole, toward which the north-seeking arrow of a compass points when free from local magnetic influence.

7) GEET Engine: an engine which can run on a mixture of water, coffee with cream and sugar, and a small amount of gasoline, or any other plasma vapor that has been derived by the GEET reactor.

8) GEET Reactor: A tube used to exchange heat inside of the chamber wherein spins a rod, which causes the vapor acceleration during the heat exchange process. This causes a breakdown of any vaporized fuel or combination of substances down to its lowest atomic level, also known as plasma.

9) GEET Rod: An accelerating rod used to create plasma.

10) GEET Gas: A plasma vapor coming out of the GEET reactor chamber, used to replace any combustible fuel as a source of heat or power.

11) Plasma: Energy which holds elements together.

12) Plasmoids: Erratic electrical particles of very high voltage emitted outside the reactor chamber under certain conditions of operation.

13) Politically Correct: Something that does not offend the oil companies or governments.

14) Reaction: When heat is absorbed with extra particles of energy drawn in changing the elemental structure of incoming heavy elements to lighter elements.

15) Rod North: In a horizontal position, the north or cupped end of the rod needs to be pointed magnetic north for optimum performance. When the incoming vapor is fed to the reactor and the rod is in a vertical position, or biblical north, it does not matter what direction the vapor is going. North will always be up.

16) Transmutation: When an element is changed or broken down into lighter elements and the mass weights of the lighter elements are always equal to the original element mass weight.

The Early Years

Many of my inventions and discoveries date back to 1954 - 1955. I would wake up from dreams and know how to build things that would work. I found this to be very interesting. I can be asleep for less than an hour, yet in my dream, I had lived for a week. During the hour of dreaming, I could build hundreds of items, which did not work. For some reason though, I would wake up whenever I found something that did work. I would draw pictures of it and explain why it worked. My mom came into my room one day when I was only five years old, at about 1 o'clock in the morning, and asked me, "What are you drawing?"

I explained I was drawing the pictures of how to do a coma surgery. I told my mother there are different types of comas and the type would determine the kind of surgery required. Most of the surgeries for comas could be minimalized to about a 30 to 45 minute operation. As I was describing a laser pointer to my mother, she had an unusual look on her face. She asked "What is a laser pointer?" I explained it was a penlight device with three buttons on the side. It would need to have a red light, a blue light, and a green light. They would light up a human torso bust of glass in the surgery room to show all of the students looking from above what part of the body you were working upon. The red is the blood stream, the blue is electrical, and the green showed where there was damage or a spinal break which caused a coma or a crippling effect due to a disconnected or damaged area. I then told her I was doing the teaching in the surgery room, but I was an old man. I described that I had a very small amount of pure white hair on both sides of my head. I also had on thick glasses and my name badge. The beautiful glass bust across the room had a gold or brass plaque on it and said that the 3M Corporation manufactured it for Dr. Paul Pantone.

Mom taught me from first-hand experience the seven types of dreams. She felt it was important for me to know each type dream to differentiate the outcomes I could expect. It took many years before I could finally identify each type of dream upon awakening. This will lead to another book focusing on dreams and their implications.

Teaching And Lessons

Looking back at this point, I have to laugh. I quit school in ninth grade, but others have introduced me, on stage and off, as Dr. Pantone. About 25 percent of my students hold PhDs, MDs, and other high credentials. At first I would explain things in my classes by saying things like, "The vapor was pulled in a radial fashion, longitudinally." When I realized the children did not understand what I was saying nor did the scientists, I rephrased my presentation so everyone could understand what I was saying.

It became apparent to me to be a good teacher, you must be able to read all of your students and you must treat them all equally. You must value the questions from the children as much as you do the

PhDs. The true challenge to me of being a teacher is teaching someone who does not speak a word of English. I show them how to build my inventions with just pictures and hand props. It is very gratifying to receive emails from people around the world who have built my invention, all with different levels of success. However, on the Internet, it only takes one person to say something bad to hurt a lot of your business, as well as damage your reputation.

One such professor had written several emails demanding I give him the plans for his car or he would tell the world how the technology did not work. Not to be intimidated, I told him he is welcome to buy the GEET plans. The negative email that Professor Shields posted is not true. He knows it, yet he was so offended because I refused to GIVE him the plans. Since I would not give him the plans free, he was willing to attack me.

If you ask some people why there are such people on the Internet, they will say it is the government. No conspiracy here. Some people will fight for their limited knowledge, thinking they are correct, since this is what school teaches. We have to allow them their personal feelings. It may be hurtful to us, but they think they are correct. Allow them their belief.

How many inventors before me have been persecuted, jailed, or ridiculed? Not to mention there are many inventors today who are terrified of government actions, the criticism of scientists, and others who fear change. It seems rather funny in 100 years, we can go from a telephone you had to crank to flip phones and watches you can talk into. You do not even have to dial the phone to call home, yet we have cars today which get no better mileage than they did in the early 1900s.

Just go on the Internet and type in "cars that can't be sold in the U.S." You will be very surprised. Some were not safe, but most are on the list because they get too much mileage.

Recently, the AMA was attacking Dr. Oz. They do not want him talking on TV about alternative methods for treatments. I do not blame them. Do you know how much money they will lose if people did not go to doctors for "treatments?" They are very concerned for their member doctors, and NOT for the public. It is sad that I do not feel I can take on all of the battles, but I choose the battles I know I will win.

I must agree some people would be in danger from the charlatans who make wild claims of cures. I have seen too many companies who think more is better. They do not do sufficient testing. The results can be devastating. I have argued for years against most uses of silver as an internal medicine, yet when told I had a MERSA infection, I did use the Colloidal Silver as a topical application, and the sore was gone and completely healed within 48 hours. However, I will not take it internally. I wish to test things one at a time.

This Is Progress?

I am fully aware that GEET will fit every piece of fuel-driven equipment on the planet. I offered in 1985, on television in Sacramento, California, if I could just have a dollar for every car that wanted to use my technology, I would probably never be able to spend all the money. The sad part is not one car company in the world is willing to reduce the cost of building their cars to become more efficient, but will spend much more to make sure we do not get better mileage. Let us not dwell on the things we cannot change; let us focus on the things we can. Knowledge is freedom.

I remember back in third grade that my teacher had a bell jar as part of our science class. She put an alarm clock onto the base after winding it up and set it to go off in 10 minutes. She put the glass bell jar over the container and clock. She turned the vacuum pump on and let us all watch. Ten minutes later, we could see the striker hitting the bells as it wildly went back and forth, but there was no sound. Sound does not travel in a vacuum. Yet, going back to current schoolbooks, no such test explains to the class why this happens. Back in the 1960s, when filming *Star Trek*, they had no sound for the movement of the *Enterprise*. They knew people wanted sound, so they made a soundtrack for ships moving in a vacuum. It was for entertainment, not something for people to believe. Today, much of the public thinks there are sounds heard in outer space because they heard it in movies. Sad, isn't it?

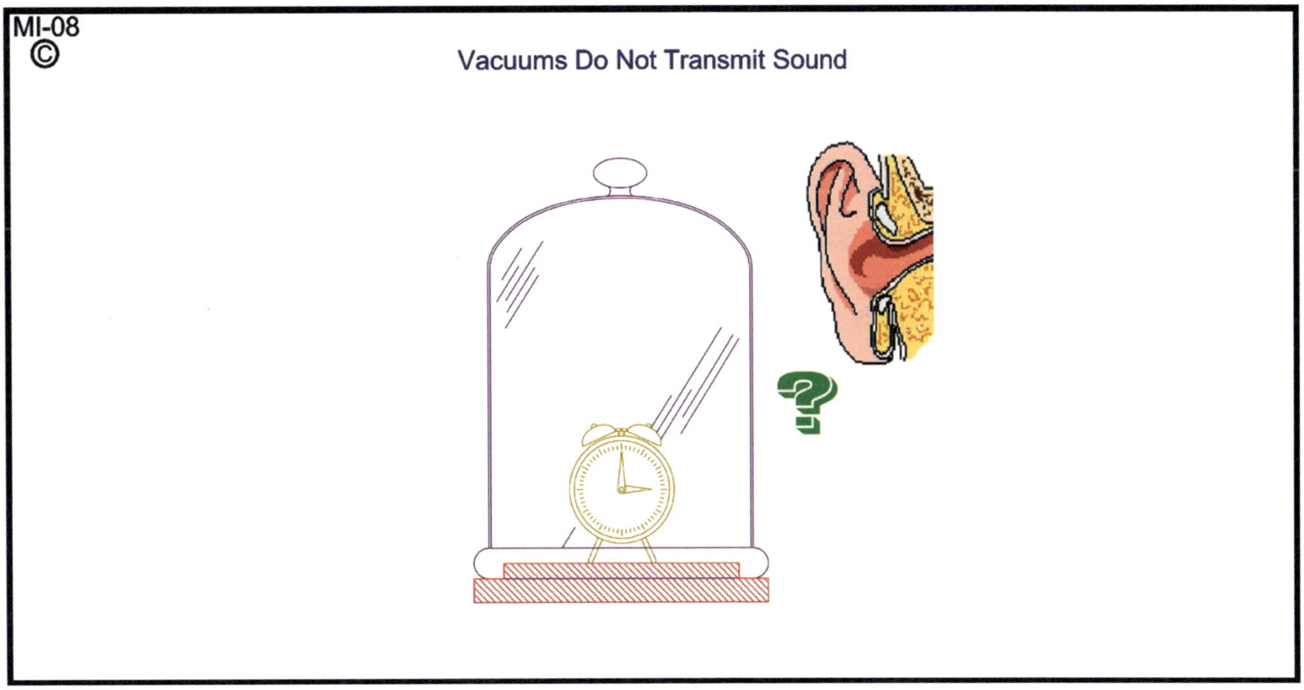

I remember back in 1957, somebody asked me to look up Nikola Tesla. The encyclopedia I used was probably seven years old, but it had about 100 pages on Tesla. There was also a very large section on the American Vacuum Society. Years later, I never see any reference to this group. It did not make sense why one of the fastest growing groups of people in the world would just disappear. As you read this book, you will begin to understand why the American Vacuum Society vanished.

Many inventors over the last thirty-five years have come to me and asked, "Why are you still trying to help mankind?" Several of these inventors met with a wall of resistance that was not understandable to them or their families. A few of them decided to bury their inventions and just have a regular life and to hell with humanity. Others felt compelled to fight the dragon, and regrettably, many of them are dead. There have been many lists on the Internet showing inventors who have things which work and who often face persecution or mysteriously disappear.

Several inventors called me up and told me they felt the pressure was coming and they wanted to know what to do. I told them before you go opening your mouth on telephone calls or the Internet, you had better have a bigger hammer than the government. One tactic the government uses is they simply remove you from the tax rolls, then come in and take everything. This includes your bank account, all your tax records, and you cannot fight back. Once you get out of jail, you most likely will do anything the government tells you in order to keep your freedom. (Safety hint: Keep a copy of each tax return in a separate location that the government does not know about).

Yes, every time I think I have seen it all, something new comes in. In an effort to stay politically correct, many of the inventions we see have to stay within the critical political court of being correct. It is disappointing most scientists I have met admitted they have seen many things that worked, but to keep their jobs, they had to remain silent. This is not the case for just the United States. Any country that uses fuel or uses the U.S. petrodollar as their base are tied to the same values, or lack of them.

We live in a world of change, and we must learn to accept it as it happens or we fight it. History teaches us that after we have revolted, many new technologies will come forward and we will embrace them. After 120 years, it is now time to embrace better technologies. We want some good mileage, clean exhaust, and to put the oil companies where they belong - on the bottom of the food chain. Let's get onto plasma physics.

What is plasma physics? Plasma is energy. It is a very high form of energy and generated under heat and pressure and by adding electrons. It can generate itself from hot and cold moving in opposite directions in a vacuum. The vacuum part requires much less energy and you get more energy out of it. Nikola Tesla did many experiments and prototypes with electrical energy that would have been magnified 10, 20 or more times had he used a vacuum.

What he really needed was to find a way to be politically correct and put meters and wires on everything so J.P. Morgan and other wealthy people could make money every time you need any electricity, water, or any other service. Most people are not aware of Rockefeller's plight in Los Angeles with the discovery of oil in Signal Hill. Back in the 1890s, a company called Day and Night made solar hot water heaters, especially in the Los Angeles area. To eliminate your competition, you simply buy the company and make the product you want. Rockefeller did this and put in pipe systems with meters. He gave a good price for new heating fuels so people did not have to worry about sunlight. He became very popular and soon all of those solar hot water systems were gone.

Was it an intentional act, or was it accidental that stronger steel came from plants where the cooling area pointed north and south? In the United States, only one company had a steel plant, properly aligned to the north and south, giving it the hardest steel. That plant was Geneva Steel. It came down to make way for great improvements in the Provo, Utah area.

As I watched it being torn down, I did mention to a few people in Utah the value of that alignment, but it fell on deaf ears. It was sad to watch the building fall just so they could build a new industrial park in its place. We call this progress?

In 1999, there was a huge chimney, in Murray, Utah. It was contaminated with products and elements dangerous to humans. They dropped the stack onto the ground and contaminated the air for a few miles.

The company who dropped it filed bankruptcy just a few days later. The public never questioned how the architects could have designed and completed the plans for the new hospital and shopping center in just a few days. However, I could see somebody planned this for a long time. Are they just blind, or don't they care?

In 1800s and early 1900s, to become a machinist or blacksmith or gun maker, a person would spend years of his life serving as an apprentice to an artist in his field. Now we no longer seek artisans. We seek people who will perform a task without knowing the overall outcome and call it progress. By incubating new ideas through their new work isolation system, the ideas can be stopped at any point in the process and diverted or squashed, if problematic, so the final product will never get to the world.

Rather than going back in time, we must change the way things are done today to become more efficient. One person should guide a project to oversee the final outcome. Today's engineers are specialized to the field they are working with no peripheral vision to the other parts of their project. This causes a major handicap. One television show explained how big corporations dealing with auto manufacturing have different engineers for the aerodynamics, the engine, the body, and even the interior components. If you are an engineer working on the body, you do not dare step out of line to discuss the wheels or the engine, or you might not have a job. As a society, we have taken this to the next level.

Over the last 50 years, it has become more important to describe people with job titles than what work they do. Take for example Walmart. Thirty years ago, you had a stock boy. Today Walmart defines the person in the same job position as an associate, or maybe an associate manager. It is all about our ego. If we can set aside the ego, most people can be open-minded.

In 1994, when I first started attending conferences for science, scientists and engineers would hound me. They would all give me their personal list of tests that could be, or should be, required to make GEET a success. Some said I needed to do the stoichiometric calculations first. For others it was a dynamometer test. Most people made these suggestions because this is all they knew.

I decided to challenge all of the great minds who were telling me how to make my invention better and how to test it, by preparing a list of thirty-seven tests that needed to be done. During one of the science congresses I went to, I passed this list out to all the attendees and I asked them to review it to see if I had missed any tests which they thought were important. What I failed to tell them was in the very middle of the page was a fake test I completely made up. It does not exist anywhere! However, it sounded very impressive. Over 650 people attended the congress and many more meetings after that, and not one ever asked me about the imaginary test procedure. I found a way to shut up the scientists by challenging their ego.

Many theorists have approached me over the years who claim that before you can have a new invention accepted by science, you must have the proper theory behind it. It is because this is the way things are done. I dropped out of ninth grade, because in my world, you simply built what you need to get the job done. There is no need to look it up in a book to see if you can do it. If you know it can be done, then do it!!! As Richard Bach — the famed author of Jonathan Livingston Seagull — said in Illusions, "Argue for your limitations, and they are yours."

This discovery may seem quite simple to you, but it still baffles my mind. Back in 1983, I spent many thousands of dollars to find out which surfactant was best for my applications of making fuel. I blended half water with gas, diesel, kerosene, and even crude oil, wanting to see which one would stay homogenized the longest. After weeks of testing, my final experiment was to light the surfaces of about 50 bottles of mixed fuels. The only fuel mixture that burned to the very last drop contained Dawn dish soap.

I did further testing on Dawn. My favorite test was dunking stuffed animals into crude oil and then slowly dunking them into warm water with Dawn dish soap. The Dawn broke down the crude oil and when rinsed off these animals, it left them clean. When the Exxon Valdez accident occurred in Alaska, I immediately phoned Texaco and Greenpeace to tell them that Dawn dish soap could clean the animals up and eliminate most of their suffering. To my utter amazement, there was no interest in letting this get out to the public. I had it explained to me that dying animals, suffering and covered with crude oil, would bring in many donations. My solution would stop donations.

Over the years since the Exxon Valdez spill, I have been on many radio shows and talked about Dawn dish soap. To my surprise, oil companies from Russia, Angola, and even Brazil phoned me to ask permission to use Dawn dish soap to clean up the animals and beaches. Since my goal is to help humanity, I told them all that it was all right. It is a shame America has so many nonprofits who have become corporations rather than individuals helping each other.

Many of you may not see the importance of such a small discovery as a dish soap. Think about it for a minute. If you can use some soap and dilute the fuel to get better mileage, then why are we not doing it? Some oil companies know this, but want the knowledge kept under wraps.

Products Do Not Align With Real Needs

Since I was a young man in my teens, I kept hearing about how magnetic energy would improve water quality or fuel quality. I decided to do the test myself and prove once and for all what really works and what doesn't. I did tests at my shop for about six months on fuel lines. I replicated the tests in Virginia, New York, California, and Australia. It took six months, but I found the answers. Magnets, when properly installed, will improve the quality of the air or liquid passing through its field, but the field must be facing north (either biblical or magnetic north).

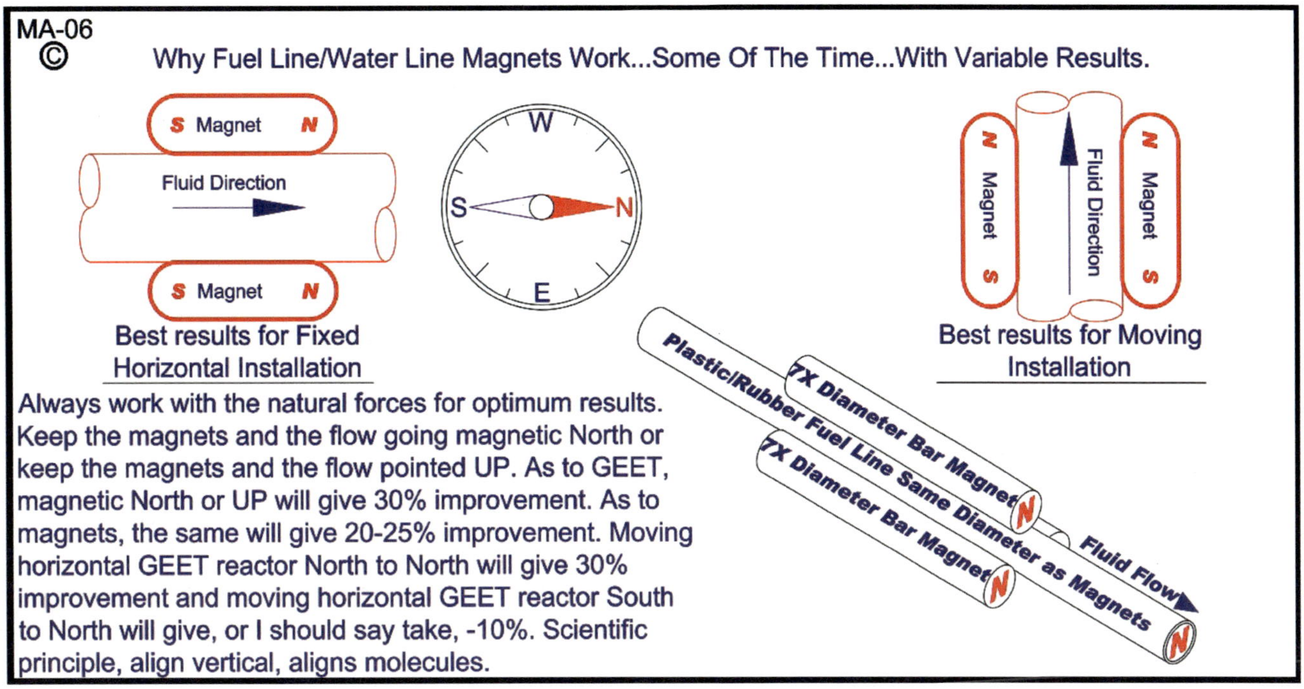

I contacted many magnet sale groups to share this information and none of them seemed to care. They make so much money now they could care less about you being a happy customer, just their sales and immediate profits.

We have many political groups who claim they are there for the betterment of humanity. I beg to disagree. One such organization of "Concerned Scientists" is in Pueblo, Colorado. They claim they wish to reduce the pollution levels from all large trucks. It only took one call to find out it was a lobbying group set up as a nonprofit to promote only the items under their control in the trucking industry. This happens in almost every field. I hope someday people will wake up and realize we need review boards made up of regular people, and not politicians, to stop the abuse of nonprofit organizations.

If you try to sue or challenge any psychiatrist, a board of psychiatrists decides if one of their own is right and you are wrong. If you take action against a police officer, the other police officers review the case. The police officer in question is almost always right. This is true in every profession. If you go after an attorney, the Bar Association reviews its fellow attorneys. Now in places like Utah, where I filed a complaint against the judge for his criminal actions to the ethics committee, I did not know he was the head of the "ethics committee." He and his fellow judges determined any criminal act by a judge is not a criminal act but it is a new law!!! How convenient for the crooks to be judged by their peers.

This does not happen just in America. This is worldwide, but it still leaves me scratching my head as to how the system really works. Here in America we shut down a lot of manufacturing so China, Korea, and Japan can make the parts at a much lower costs and higher profits. When I ordered a Wisconsin Robin generator which was made in Korea, it cost me more than one made in America, and the quality was completely missing.

In 1983, when I began building my first few fuel systems, I used Permatex aircraft quality sealants. By 1985, I noticed the same aircraft quality sealants were no longer holding up in the field, especially if I use gasoline containing alcohol.

When examining some of the new containers next to the old containers of Permatex, I found the new products dissolved in gasoline with alcohol. Maybe someone out there can explain to me why you would make a gasket sealant, which would dissolve in alcohol after you put alcohol in the petroleum. This would definitely convince anybody with a carburetor that they would want to switch from a carburetor to a fuel injected system, even though you cannot calibrate fuel injectors and they are very inefficient. The proof of this is very simple. Look at the mileage we get today on new cars with injectors compared to the mileage we used to get in the 1920s. If injectors were so perfect, don't you think the mileage would have gone up a little?

We tested about 80 injectors and giving the same pulse to each one, we got as much as a 50 percent variable on the amount of fuel delivered from each one of these identical injectors. If injectors were properly engineered so they could be adjustable, all of your injectors would be matched for volume of fuel delivered to each cylinder. This would give you better mileage. Why is this not being done?

During a recent radio show with Jay Thomas, a caller from Ohio stated the engines he works on at the assembly line only get about 30 miles to the gallon. He did say the same engine was sold in Europe with different computer controls and they get almost double the mileage. When will the people wake up and see congressional "representatives" and senators are not working for us, but for the oil companies? Maybe if they represent us, we could get the same mileage as our European friends.

Where Is The Justice?

If you took the time to look at the 1951 Invention Secrecy Act, it would alarm you. This act granted the Patent and Trademark Office the official duty of notifying federal agencies of any new invention, which might interfere with national security issues. At least, this is what it says. When I questioned why it was only revised once in 1971, I was told the public is not to be told of the newly revised national security issues. Is this because there would no longer be security?

Any inventor who has gone through what I have will understand big business does control what the public gets. This will become very apparent, as happened to me, the first time you go into a car dealer and offer to buy a car and they notify you that you are trespassing. They said "If you come back, you will be arrested!" What is even worse is if you try going to a wrecking yard to get parts and you are told they cannot sell you anything or they will be harassed and put out of business.

I can go on for days of all the criminal things that have been done against me and how government agencies make it look like I'm the bad guy. In one court case, Detective King, under penalty of perjury, claimed I had admitted guilt to a criminal act I did not do. When it was my turn to take the stand, I offered the judge a copy of the original recording with this detective proving she had lied while under oath. I also had a certified copy of the tape in transcript form.

That judge should have removed himself immediately, but he repeatedly violated his duties and in front of a jury for three and a half days kept saying, "When you find Mr. Pantone guilty..." Even with the cards stacked against me and none of my evidence accepted, I still won. It is hard to believe had I not seen it myself and heard it with my own ears, the number of crooked judges we have in this country. Maybe they do not realize with their criminal actions, they dishonor the very robe they wear. When doing a personal check on one judge in Utah, I came across a group of companies run out of the courthouse by judges and operated by a relative. This would explain why everyone, even those who are innocent of all charges, need to pay for court costs.

The Rod Is Where It All Happens

What is a GEET reactor rod? The reactor rod is a rod within a heat exchanger used to accelerate the fuel vapor and air mixture to break down the fuel to the lowest atomic level. A smooth rod delivers better performance than a rough rod, but the rod itself does not have to be polished or machined to perfection. That is unless you are doing it for yourself, you own a machine shop, and you do not mind spending the time.

It may be desirable to put cavitation divots on the rod to enhance the performance, but it is yet to be done. Most all other possibilities have been tried and have not been successful. Even plating the rod has not shown it can return the costs incurred.

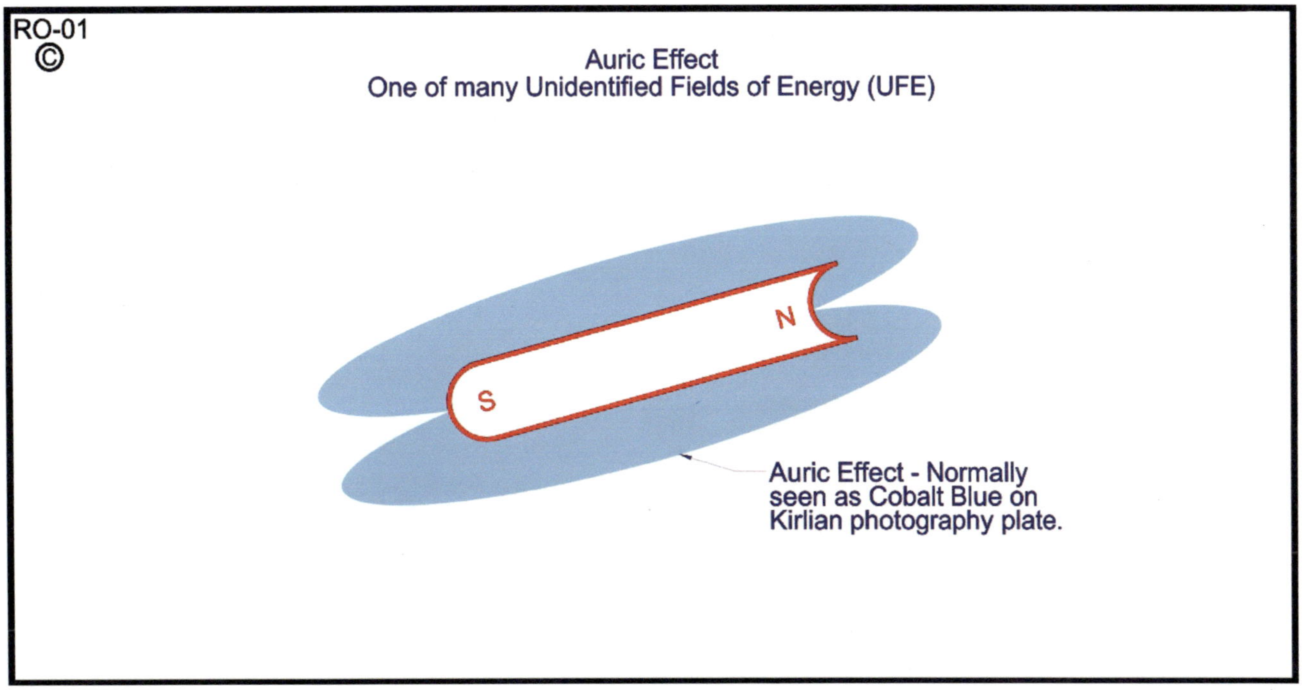

What is the best material to use for a rod? We have tried glass, hard plastics, stainless steel, steel, aluminum, brass, etc. They all work. Cold rolled steel is about the cheapest and easiest to work with, so this is what we usually use. As you go through different sections of this book, you may choose glass, or even a gold-plated rod. The strength of the magnetic field that is self-generated around the rod may be more or less intense, depending on the material and the fuel vapor you are using. One of the most unusual phenomenon we are seeing is a cooling or heating effect when using a digital handheld thermometer.

There is something occurring as the engine exhaust leaves the final exhaust pipe which causes a physical cooling upon the exhaust pipe. Within the center of the reactor is a hot spot, sometimes twice as hot as exhaust coming out of the engine. When we have tried to validate these temperature readings using thermal couplers or computer analysis units, the readings of the actual temperature is different from when we used a handheld digital thermometer. We have also tested the wire temperature when electricity is pulled from the battery to the starter of a car; although that wire can get very hot, it does have a cooling effect at the post of the battery.

During one class, a professor asked, "Why is there ice where you say the hot part of the reactor is?" As I looked at the picture, I saw the ice and could not answer at first. Then I knew I was getting a false digital hot spot reading and it showed me where the spinning rod was located in the reactor.

Although the free plans show a bubbler to make the vapor from mixtures such as soda pop and gasoline, I do not recommend doing that for any commercial application or for automotive use. Each type of fuel requires a different length and diameter of rod based on specific formulas. Soda pop would use a rod around two inches long to achieve atomic breakdown. Gasoline requires a three and five-eighth inch rod length to achieve its atomic breakdown. Each fuel or blend of fuel requires a different specific rod length to gain optimum performance.

To begin with, let's clarify things found on the Internet. When we talked about using soda pop, coffee, pickle juice, and other fun fuel, it was solely for entertainment. The only reason we ever talked about these kinds of fuels, rather than gasoline, diesel, and kerosene, was to get television coverage. If we call the TV station up, talk to their news department and tell them we could get double or triple the mileage, or even a measly fifty percent increase in mileage, they would not allow it on the air. It is not newsworthy.

However, the minute you call up and say there is a crazy inventor out here running an engine on soda pop, all the news stations show up. The problem is they want you to be the clown of the news hour. Some of the stations have even posted captions next to my name such as, "See the crazy inventor Paul Pantone." Most of these reporters do not have a clue if the engine will run on soda pop.

Back in 1983, we welded tabs at each end of the rod and then filed them down to help hold the rod in the center of the flow area. It was many years later we discovered that fastening the rod to the reaction chamber caused it to short out and stop other phenomenon from occurring, including plasma. Although we made many discoveries with the tips welded on, they were not relevant to a full plasma reaction and restricted the use to only that of a vaporizer. The rod in a GEET needs to move freely while spinning so that the rod will stay in the center of the electrical gravitational field it creates when making plasma.

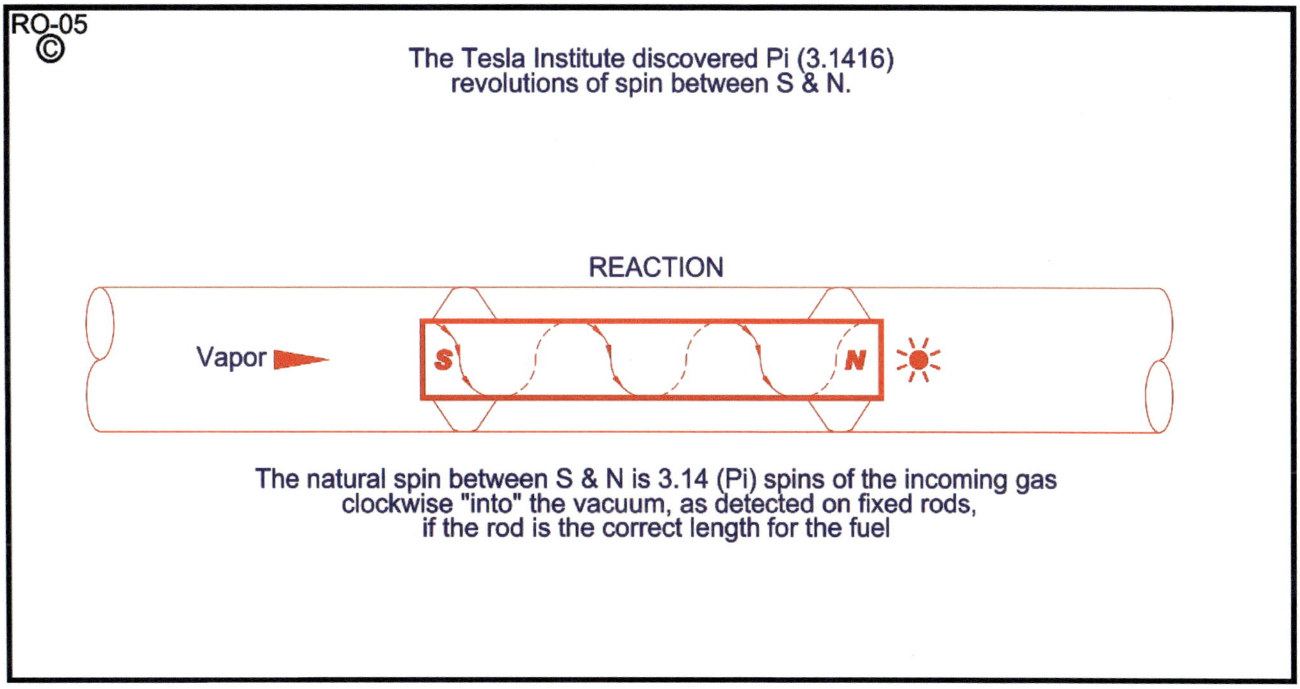

The original rods were cut 90° from the rod surface and both ends look the same. This later proved to be a problem when people would take the units apart to try and find the secret of what made it work. Then they would put the rod back in and it would be backwards. When the rod is inserted backwards, it has a tendency of not allowing a plasma reaction, but only a vapor reaction.

The evolution of the rod to its current shape was made over a ten-year period between 1994 and 2004. At first, the rods were square-ended. In 1999, we concaved the north end of the rod. This is the part of the rod closest to the exhaust or heat source of the engine that causes the plasma generation. Rods prior to 1994 normally had tips welded on each end and rounded down to help center the rod to the chamber. These welded-on tips restricted the spinning of the rod and grounded the rod out.

In 1999, the concave end was placed to the north. This allowed the rod to move back to the center of the magnetic field it generates more quickly. We need more research on the airflow dynamics of the rod cup at the end. It seems to propel the unit into the wind or into the flow. This design might be used to improve bullet trains, trucks and other equipment, as well as certain projectiles. They could possibly achieve a fifteen percent better distance or range with the same propelling energy.

The bullet shape to the south end causes the airflow or fuel vapor going into the air chamber around the reactor to be more uniform, allowing the rod to spin freely within the chamber. It is this 'fitting action' of the spinning metal rod and the air moving around it, which cause a type of electromagnetic/gravitational field. It is very similar to that of the magma at the center of our planet.

When the rod is the wrong length, it will show a very unhealthy aura or auric fields. This can be detected with Kirlian photography, or a simple seventy-nine cent compass. The class training manual books on page RO12 show charts that indicate the length of the rod and the clearance of the rod within the chamber. All of this information is in generic form and works. Although it is not a perfect formula of how to size a rod, they will adequately suffice to get an engine to run with approximately ninety percent less pollution. The formulas mentioned in this book are only needed if one is disposing of a waste product that needs to be completely changed to a neutral state.

RO-12
©

Rod Length By Orientation

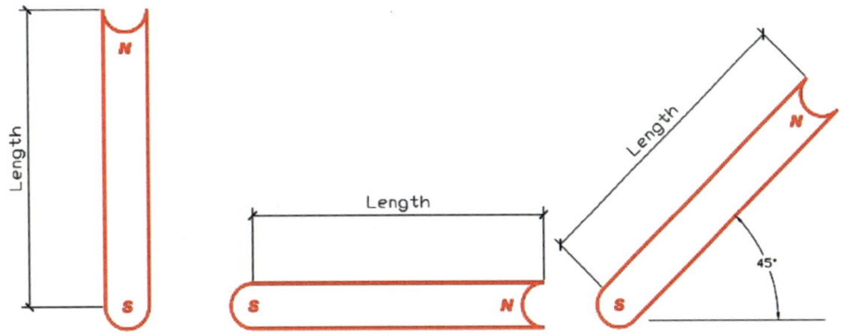

Rod Clearance by Fuel Type:

Small chain fuels require smaller clearances

Large chain fuels require larger clearances

FUEL	#1	#2	#3	Rod Clearance by Fuel Type
Gasoline	3.625" [92mm]	7.25" [185mm]	5.4375" [138mm]	0.06250" [1.6mm]
Diesel	4.5" [115mm]	9" [229mm]	6.75" [171mm]	0.09375" [2.4mm]
Crude Oil	6" [152mm]	12" [305mm]	9" [229mm]	0.12500" [3.2mm]

We now have a rod-reader which will soon be available to all. It will chart on the computer the magnetic field of the rod detected by the reader so the operator can make the proper rod within 1 mm length to get a perfect burn.

The Free Plans

In 1999, the Free Plans for Small Engines was donated to humankind for educational purposes to allow every person on the planet to build one unit for personal use ONLY, but not to be used for any commercial application. When I made this donation, it was because OPEC representatives were coming to my home to make an offer to "take over the technology." I felt the need to get the technology out to humanity before they put it up on a shelf to be forgotten forever. With this in mind, I took the worst plans I had to show proof the concept did work.

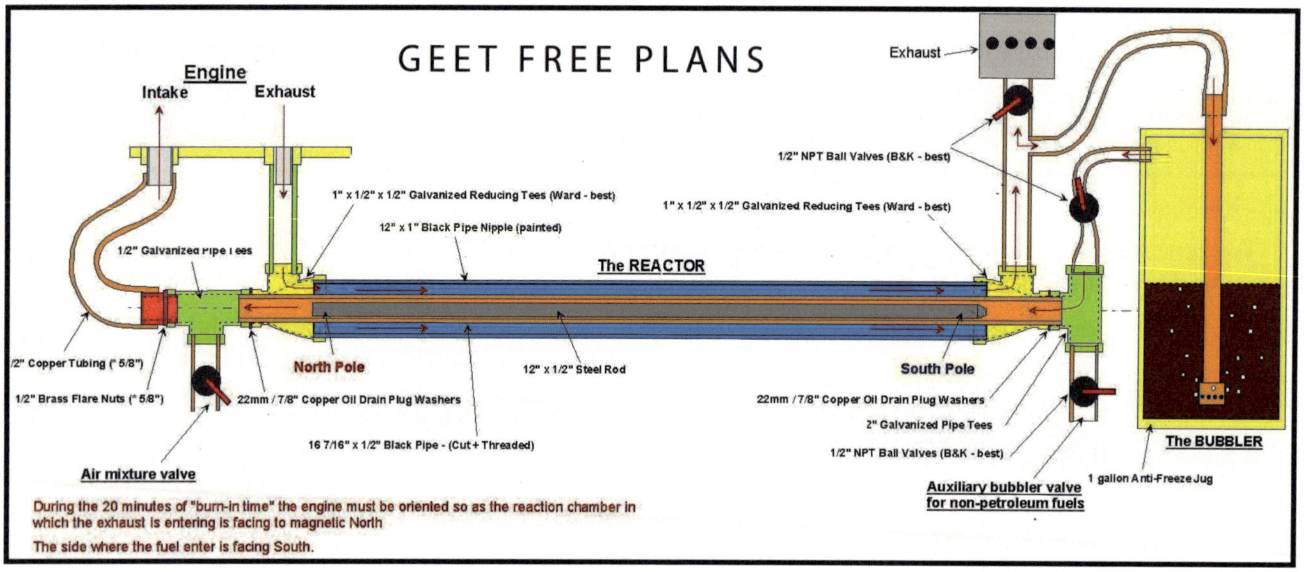

Within days, people around the world were downloading the free plans and saying thank you. However, for every thank you, I had at least two negative comments. Some people claimed they did not have the ability to screw two pieces of pipe together and wanted me to supply them completed engines, classroom supplies, generators, and even kits for their cars. Many organizations took my free plans, posted them on their sites, and asked for donations in their name so that their foundations and organizations could continue to give you the best knowledge. This is how our educational system works.

Some boasted of their success in building GEETs, and years later, we found out they did nothing but lie and use this lie to raise money so they could stay home and party all day. Nevertheless, there was a good side to these donations as letters, videos, and emails poured in filled with exciting details of how well and how clean their engines ran.

One not-for-profit raised a great deal of money in Australia, claiming they were working with me. At that time I was detained for three and a half years, and they did not even have a working unit! If you sent a donation, it went to partying.

My contribution was only for small engines up to 20 horsepower. People violated my free plan stipulations and efforts from day one. Many people began installing it on big trucks, ships, motorcycles, and even a helicopter with no intent of making any donation or paying any royalties. These people felt they could hide in the woodwork and I would not know what they were doing, but they were very wrong. At the bottom of the drawing, I had asked that anyone who got these plans and produced one that ran please donate what they could afford to help the inventor out so I could get more information out to mankind quickly. I am sorry to say that I only received $105 from one person in America and one person in Europe. Later, I talked to many of the people who took my plans and put them on their websites seeking donations in their own names. Some boasted of raising thousands of dollars on their site. Now with copyrights in place, I will get them to remove any links seeking funds for their private clubs by using my technology.

It is hard to want to help people when the only response from them is to 'give them more!' It has been 16 years since I gave away the original small plans, and still I get emails every day from people who want a free kit, engine, or generator. Maybe times have changed so much that I have lost touch with reality. When I was growing up, if you wanted something, you went out and earned it. If somebody gives you free plans to build something and it works, you send them something, even if it is just a thank you letter.

The people who responded in a positive way gave me some encouragement that there were folks out there who really do care about this planet. I must admit most of the responses were purely from people who want to save money. I have no problem with that since I know it will help clean up the planet. In the giveaway of the original plans, I told people to use a 12-inch ferrous metal rod, with detailed instructions on how to weld on the three little tabs to each end so you could center the rod while it was running. These tabs on each end make it a vaporizer, and you can expect to receive about a 30 percent increase in mileage. For many people, that was very beneficial enough.

Although the free plans were only for small engines, I was working on large engines which were over 500 horsepower. Some of the engines were six feet tall and 14 feet long with seven-inch pistons. The scientists and engineers who came and saw GEET working were only looking at small engines, and they assumed I was not aware the same technology could be used for big engines like ships, or huge generators. When others saw the potential, most would not say a word to me, so they could supposedly capture the market without any royalty or recognition.

If GEET Doesn't Work, How Could We Go 80,000 Miles?

One of these professors went on to do a radio show, where he mentioned how stupid the inventor really was for not realizing the technology is adaptable to big equipment. I think we all know you should not assume anything. The patent which was applied for and issued worldwide stated clearly this fuel delivery system was for any and all equipment that uses fuel. We could not get the wording to be any broader than it was. It also said the fuel could be from liquid, solid, or vapor.

Yet still there have been a few Patents issued in France and Korea, using my drawing from my original Patent, but now with claims, they invented the new technology. Really?

Although the rods we used at first were machined out, we found out this was not necessary. When I first started playing with this technology, I was having parts machined so they looked pretty, but they were very costly. It also took a long time to get custom parts made. This was an unnecessary expense to impress everyone, but when I started doing physical testing with just plain cold rolled steel from a hardware store, as opposed to beautifully machined rods, I did not see any noticeable difference. Now do not get me wrong. I said, "Noticeable difference." The equipment I was using to analyze emissions was able to read oxygen by percentage, hydrocarbon by parts per million, and CO and CO2 to the hundredth of a percent. Had this calibration equipment been much more efficient, I might have been able to read smaller particles, therefore requiring machined parts to accomplish zero emission.

On my 1984 Suburban, before modifying with GEET, we achieved around 8 to 8.2 miles per gallon. Once modified, the mileage increased from 16 miles to the gallon to a high of 45 miles to the gallon, depending on where the manual controls were set. We did this with 85,000 miles of testing as we toured the United States in an effort to show people GEET did work. During the trip, many mechanics asked if we had tried different metals and the answer was always yes. I explained the reason we said "ferrous metal" was to minimize the number of emails I might receive asking if they could use different materials.

Many of the questions I received were about test procedures, types of materials, and just general questions of what fuels we tried. Back in 1983, my goal was to have zero pollution. Then it changed to getting rid of toxic waste. Later that year, the question was if I could take crude oil right out of the ground and use it as fuel, without the cost of refining. Thanks to Texaco, Chevron, and private well owners, I was able to complete those tasks in just a few weeks. The results were phenomenal. Crude oil, right out of the ground, had 25 to 45 times more energy than gasoline with 90 percent less pollution. One batch of crude oil from Southern California contained about 80 percent saltwater and this fuel mix produced 90 times the usual mileage.

RE-15
©

What Is Really In Crude Oil?

Paraffin-Based Crude Oils: These contain higher molecular weight paraffins which are solid at room temperature, but little or no asphaltic (bituminous) matter. They can produce high-grade lubricating oils.

Asphaltic-Base Crude Oils: Contain large proportions of asphaltic matter, and little or no paraffin. Some are predominantly naphthlenes so yield a lubricating oil which is more sensitive to temperature changes than paraffin-base crudes.

Mixed-Base Crude Oils: The 'gray area" between the two types above. Both paraffins and naphthlenes are present, as well as aromatic hydrocarbons. Most crudes fit this category.

PLASMA REACTION CHAMBER

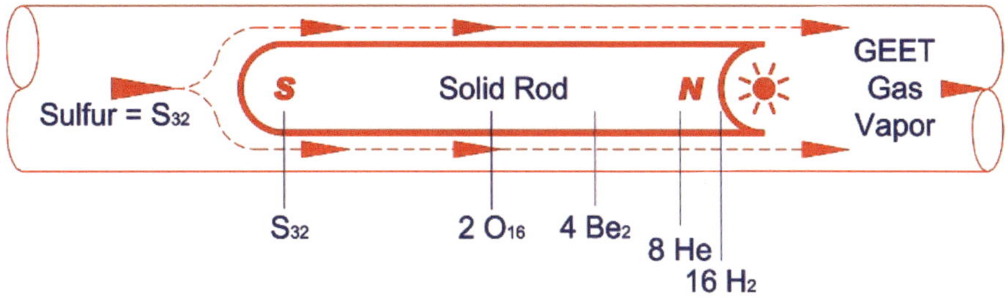

Around 1991, while driving through Richmond, California, I stopped at the Chevron research center and asked if I could get crude oil samples. They shipped them to me about a week later. I went to a California smog shop in Sacramento and videotaped the GEET running on Angola crude. The 0_2 coming out of the tailpipe was cleaner and in more abundance than what was in the ambient air. The air in the shop was showing as 5 to 7 ppm of hydrocarbon, yet the exhaust only showed one to 3 ppm. As for CO and CO_2, they were almost nonexistent.

I sent copies of this video to Chevron, Texaco, and several other places who had given me crude oil before. After that date, nobody would send me crude oil again or even answer the phones. Some of the suggestions I received for fuels were taking me further and further away from cleaning up the pollution on this planet. This is why I decided to focus on equipment that used conventional fuels to provide better mileage and less pollution.

Inventors Get No Respect

During the last 37 years, I received emails and letters from inventors or their family members from around the world. Many of these inventors had designed new types of engines, but only a few had ever taken the time to build one. Those who had built engines fell into two categories. The first category was inventions which really worked to put out good horsepower and good mileage. They were threatened and told never to release their knowledge to the public.

The other category was inventors who built their inventions and were told by many investors how many billions of dollars the technology was really worth. Sadly, this group refused all reasonable offers. They built mistrust for many wealthy companies and people who might have been able to help them get their invention to humanity. It also caused groups to put high price tags on their tremendous knowledge, far beyond what any reasonable business was willing to pay.

Rather than coming out with a new engine with a GEET on it, it really made more sense to try to retrofit every piece of equipment on the planet so the old equipment would not be obsolete. Every time I took one step forward, it felt like I went two steps back. I did a radio show in California back in the early 1990s. I mentioned that GEET could reduce emission up to 100 percent. A few weeks later, on the bottom of an agricultural bill, someone slipped in a sentence saying no fuel line could be within six inches of the exhaust pipe. This made it illegal to deliver wet fuel to the exhaust pipe. It is a good thing we do not deliver wet fuel. We only deliver air blended with a new vapor fuel that we identified as GEET Gas.

A few Hollywood motion picture people and politicians from other countries got a hold of me. It seems like all these rich people wanted one-of-a-kind cars. I told them to forget it. It did not make sense for me to put my engine on one car. I could spend the same amount of time on an engine model produced in the millions. Then again, rich people like to think they have the only one of anything, and I am not interested in that. Do not get me wrong. On occasion, I built one of something and gave it to someone as a gift. I did this because I realized there might not be potential for that product.

Let's look at it like this. How many 1930 Rolls Royce Phantoms are running around that might want better mileage? If it takes sixty hours to make ONE and there are thousands of Suburbans. Which one is the bigger market for my invention? The standard 350 cu. in. engine was very popular here in the U.S. and Europe.

Although the parts look nice when machined out, they are not worth doing. Getting past the politics can bring you to your knees or cause complete financial collapse. It was at this point I realized I must use pieces and parts anyone could get from the hardware store. The parts machined out for a car kit would cost around $1,500. This is because we were using such small quantities. If we did large volumes and could use an assembly line, you could probably get the cost down to around $250 a car. However, if you can remove the cost of the catalytic converter, smog pump, all the brackets, and not have to modify a computer system, it would reduce the current cost. It would also put out less pollution. But, this would not be politically correct.

Rod Shape And Position

Teaching people how they can put this on themselves with standard parts from any plumbing store makes it very affordable for the do-it-yourselfer. Yes, I know this upsets many machinists who want to make a lot of money operating their lathe machines. Let's look at just the rod itself for a standard small gasoline engine. The rod is about 7/16 inches in diameter x 3 5/8 inches long. You can find it at any hardware store for about three dollars for a 4-foot piece. The cost is about $.25 and spending one hour of your prep time to have a rod ready for GEET. The alternative is that you can hire a machinist for $50-$350 and get the same rod.

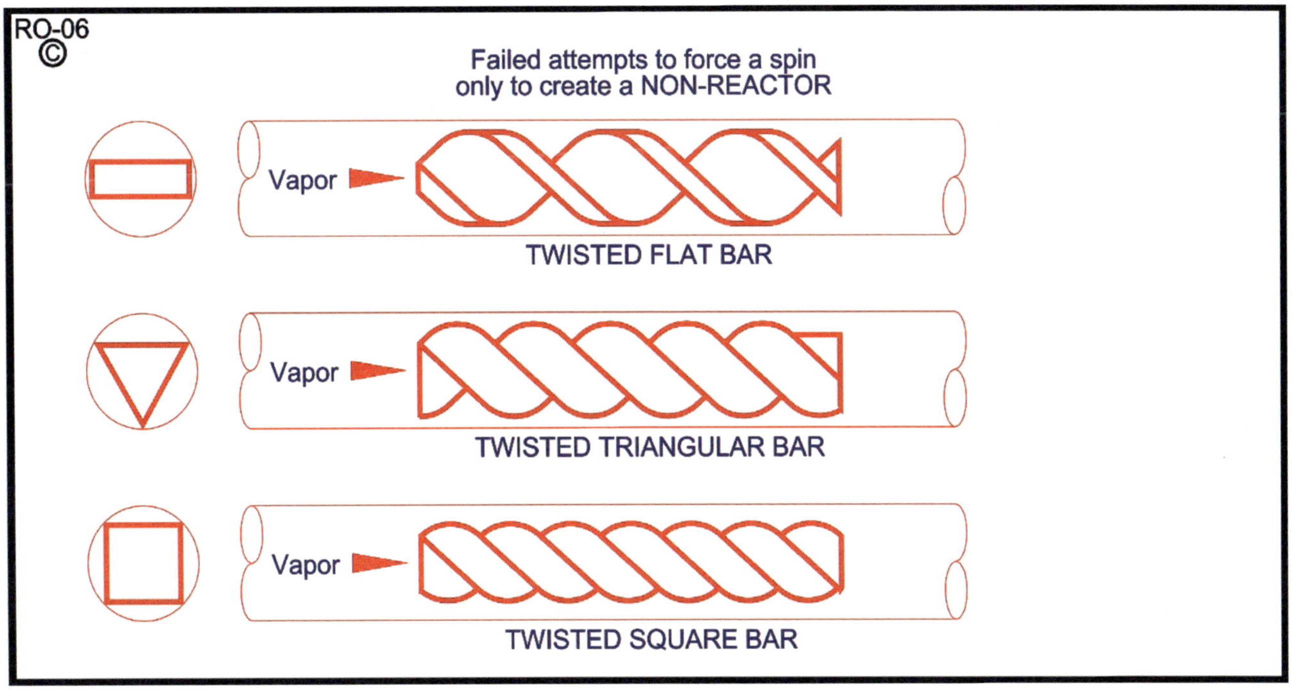

People have asked me if I have tried different shapes for the rod, such as a flat shape. Yes, I have. A flat rod will only work if the number of turns is exactly the number required for that fuel at a given speed for the fuel delivery. If you increase or decrease the speed, it will not have any efficiency. Many different configurations will work, but are only desirable if an engine is not of variable speeds and needs only one specific speed to operate. A few of the people who bought plans have spent the extra money to plate the rods with gold and silver, but this did not improve the efficiency to any noticeable degree.

For over 37 years, we have found that pi (3.14) keeps popping up in our calculations. The air entering and leaving the reaction chamber both need to be 3.14 times the diameter of the reaction chamber to allow the rod to move and stay centered in the electromagnetic gravitational field. If there is not sufficient room to regain momentum into the spiral or vortex prior to leaving the reactor accelerator, the efficiencies drop drastically. When reading the outside of the chamber during operation, we have found that an ideal temperature, using a handheld digital thermometer, is between 250 and 300°F at the engine exhaust (north) end of the rod.

If the rod is too close to the engine, it will force the rod to the furthest point it can get away from the heat. If the rod is too far from the engine, it will try to get as close to the engine as it can. In this case, the magnetic signature will not be at the top of the rod. It might be as much as one inch on the top of the rod, and the compass will show the bottom of that magnetic field to be as much as one or two inches from the bottom of the rod. To correct this problem, you must move the rod further from the engine by having a longer exhaust pipe coming to the reaction chamber.

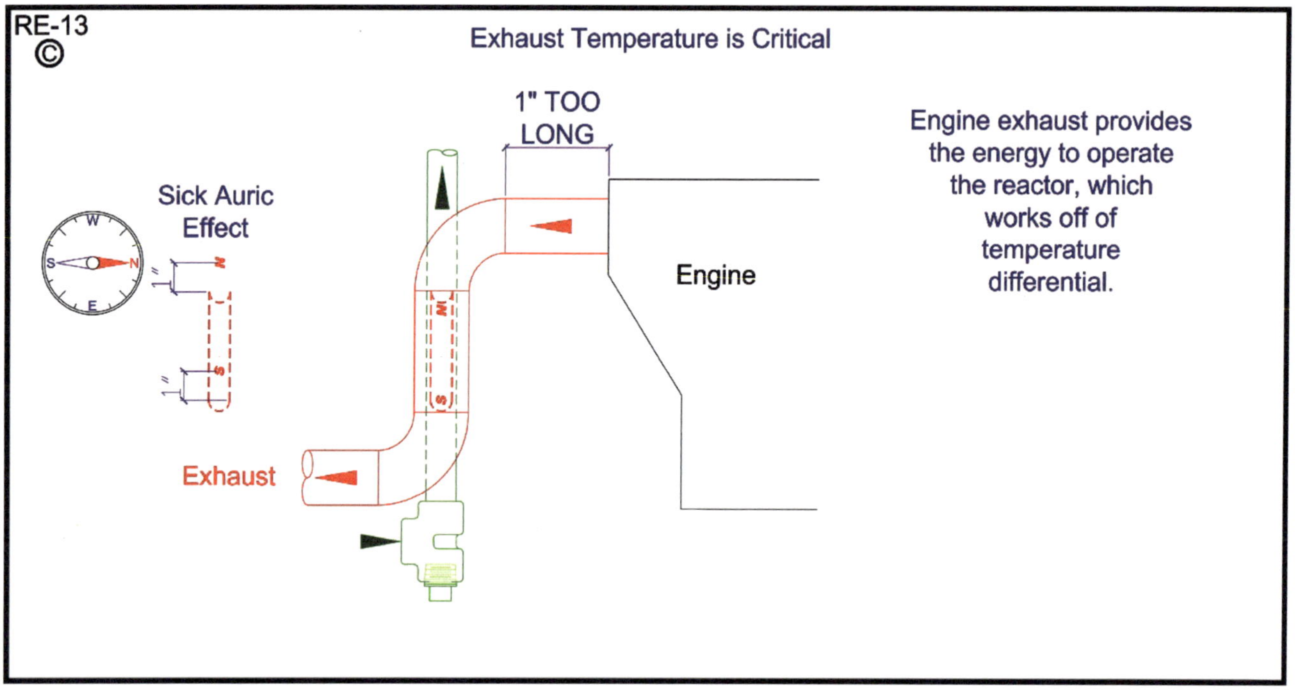

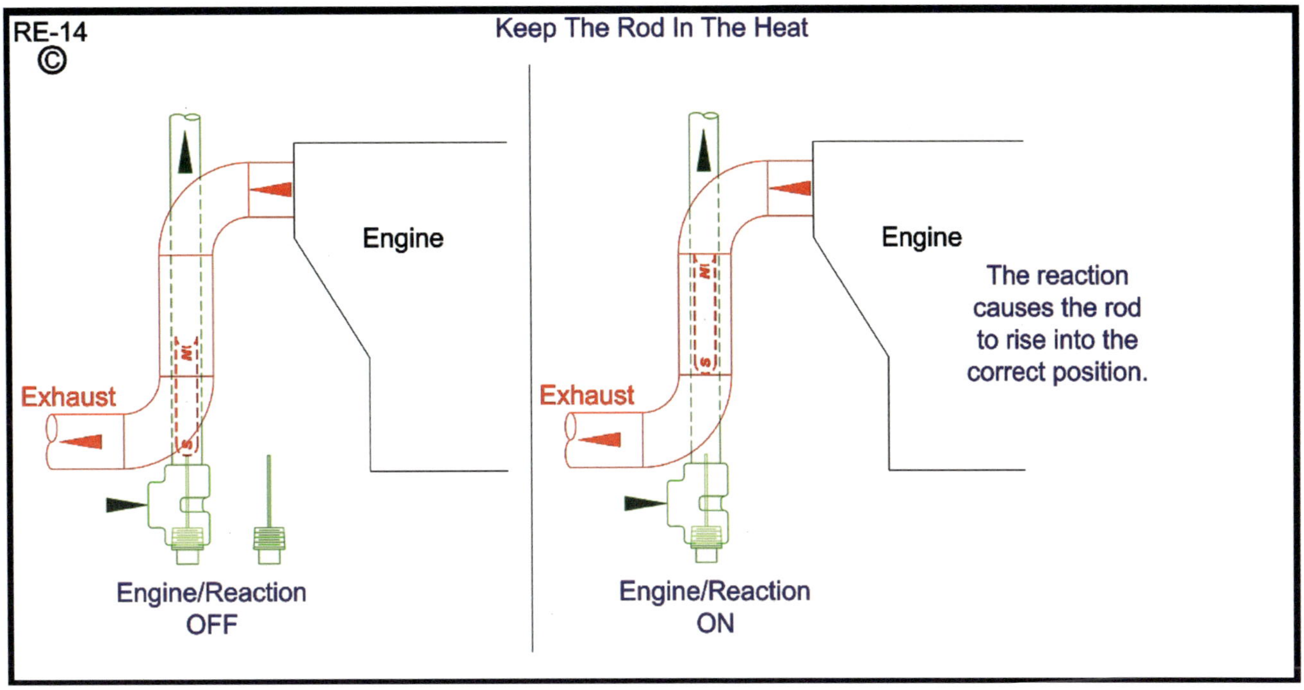

Although the rod will attract the compass needle into showing north and south on or around the metal or glass rod, it has no magnetic qualities and cannot attract even a paper clip or iron filings. Yet the intense energy detected coming from the rod at the north and towards other magnetic fields causes changes to both fields.

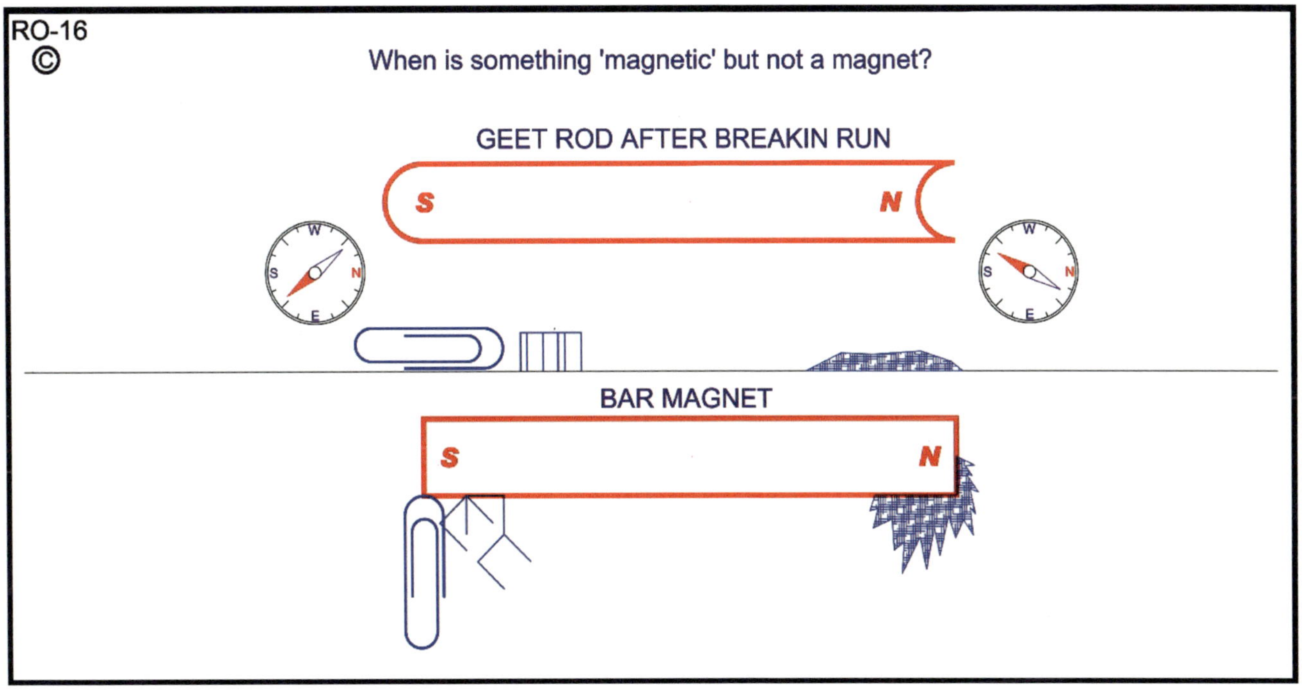

The Flying Car That Wasn't

For several years, we told people the hot spot you read with the digital thermometer indicated where the center of the rod was when the engine was fully running. It was not until late 1999 when one of our contract dealers, Dr. Paul Moller, inventor of the Skycar, phoned to tell me his GEET was not working properly on his rotary engines. Dr. Moller said the emissions had gone down drastically and his engines were no longer overheating, but he was not able to detect any temperature spike within the reaction chamber.

It was only because I had a contract with Mr. Moller that I agreed to send my technician to him. I based my agreement with him for my technology on money paid to him for R&D as well as sales. He admitted to me he had no intention of ever marketing any product. Getting money for R&D forever is great when you never have to produce anything. Preselling cars would give him a great deal of R&D capital, and he would only be obligated to deliver the cars when he said the product would be ready (i.e. out of R&D). Which would be never.

The following day I sent Mike to Davis, California, to find out why Dr. Moller's unit was not showing this hot spot that we had identified. While the rotary engine was running on a test stand, Mike checked the temperature of the exhaust with our handheld digital thermometer. He saw there was a hot spot indicated. Dr. Moller had thermal couplers connected to the exhaust pipe and they did not indicate any temperature increase from his computer screen.

Mike phoned me to inform me of this new discovery and I was surprised it took four days to find. Mike clarified they only worked a couple hours a day. He said the millions of dollars of equipment just sit covered up, and they play almost all day. Mike further validated the fact when Mr. Moller was on the cover of Popular Mechanics, it was a faked cover. Since he could not get his flying saucer to fly, a crane lifted the Skycar and the photo was altered so the public would think it really flew.

That cover made me look for Mr. Moller years ago. His receptionist said he was too busy to see me. It was funny years later when he phoned me wanting to talk. He said he was having a problem with his rotary engines running too hot. I told him GEET would stop the problem, and later that week we entered into an agreement. Now, he claims he invented this new great carburetor! However, he did mention to a reporter in Europe that it was my design.

When you reduce the weight of his flying car by the amount of a cooling system, it should fly. That is if he really wanted it to.

False Readings

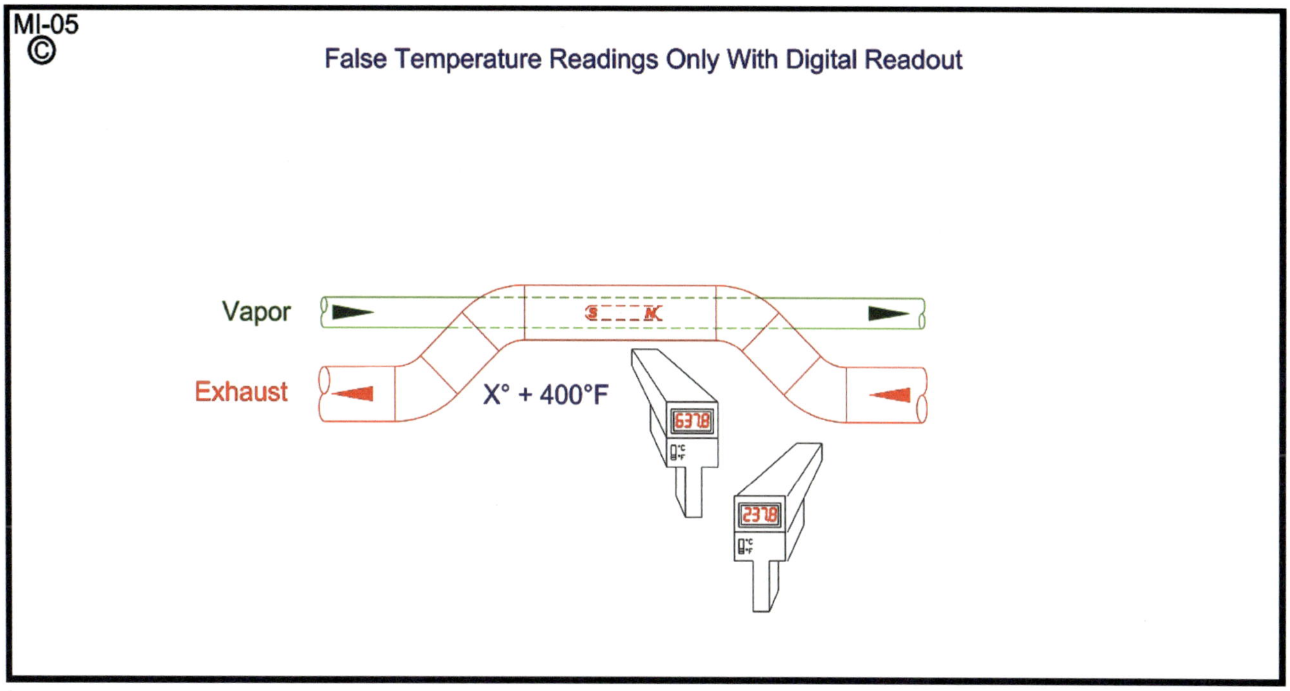

We know something was causing this false reading with the digital thermometer. Since it does indicate the location of the rod, we continued to use this device to locate the rod while the engine is running. It was at this time we realized when you have a new invention, you might have to invent the tools required to do the testing. We are doing that as you read this book. We try to keep an open mind and listen to all questions, as well as other people's opinions, to figure out what might be happening. As an aside, a few people have said they have detected a cooling as electricity is pulled through a wire to a ground. We have nicknamed this "cold power."

Media Exposure Surprise

In 2009, while doing a radio show with George Noory on Coast to Coast radio, I mentioned it is my belief there are as many female inventors as there are male. I went on to say I believe the reason most inventions are credited to males is women have men representing their technology so it would be politically correct when it got to the marketplace. To my amazement, I received hundreds of emails from female inventors who wanted me to open a female inventors club.

After careful consideration, I believe it would take a woman to run a club like this. Many are qualified to do so. A few Hollywood producers got a hold of me and said they would be happy to put a show like that on television, but at the time, I was too busy trying to promote my own company. Someone like Lori Greiner would best produce a show like this. She is an inventor on *Shark Tank*.

When doing a show on Coast to Coast, I was told they have about 3 million customers who listen and I would be lucky if I got 3 to 5 percent of them to actually respond to anything good. Most people just do not respond. You can imagine my joy when almost 1 million people went to my website during the next 24 hours and downloaded the free plans. That one radio show generated about 10,000 email requests during the next week, and I spent day and night answering emails until I caught up.

During the last 30 years, I have probably done about 700 hours of radio shows. About a third of them were just off-the-cuff where the radio hosts call me and say that someone did not call in for an interview and asked if I would take their place. All of these shows were all unrehearsed. No script, no prompts, just straight from the gut. Like anyone else who has public exposure, we always had one or two hosts who did not know what he or she was talking about. Callers would also challenge whatever was being explained. When dealing with science and new technology, you must have an open mind.

Orientation, Materials And Power

When running a GEET rod in a vertical chamber, you may have to put a small piece of metal below the rod as a spacer so the rod can operate between where the exhaust comes in and where the exhaust comes out. If there is too much area for the rod to fall below the exit, sometimes the rod cannot get into the magnetic field and perform properly. (See Drawing RE-14 previous) We have tried putting pins through the reaction chamber. It seems that everything we tried ended up with vacuum leaks.

I cannot stress this enough, for some reason Teflon tape does not work with a plasma generator. After testing every brand of sealant, we found the only sealant which would stand up to the hot and cold and vibration was Permatex Gasket Maker. It is also known as red RTV. This product is available at almost all auto parts stores. We did find when using pipe dope, if a vacuum leak occurred, it could cause an implosion destroying the reaction chamber. The red RTV is a silicone gasket maker material.

During 35 years of testing, we have seen many anomalies which caused concern. One such anomaly looked like the aurora borealis coming out through the metal at the north end of the reactor. When the engine was restarted and revved up to a high speed, far beyond the 3,600 RPM indicated on the engine, the aurora borealis affect went from the North Pole all the way to the South Pole of the rod. This was completely around the reaction chamber and lighted up the entire yard.

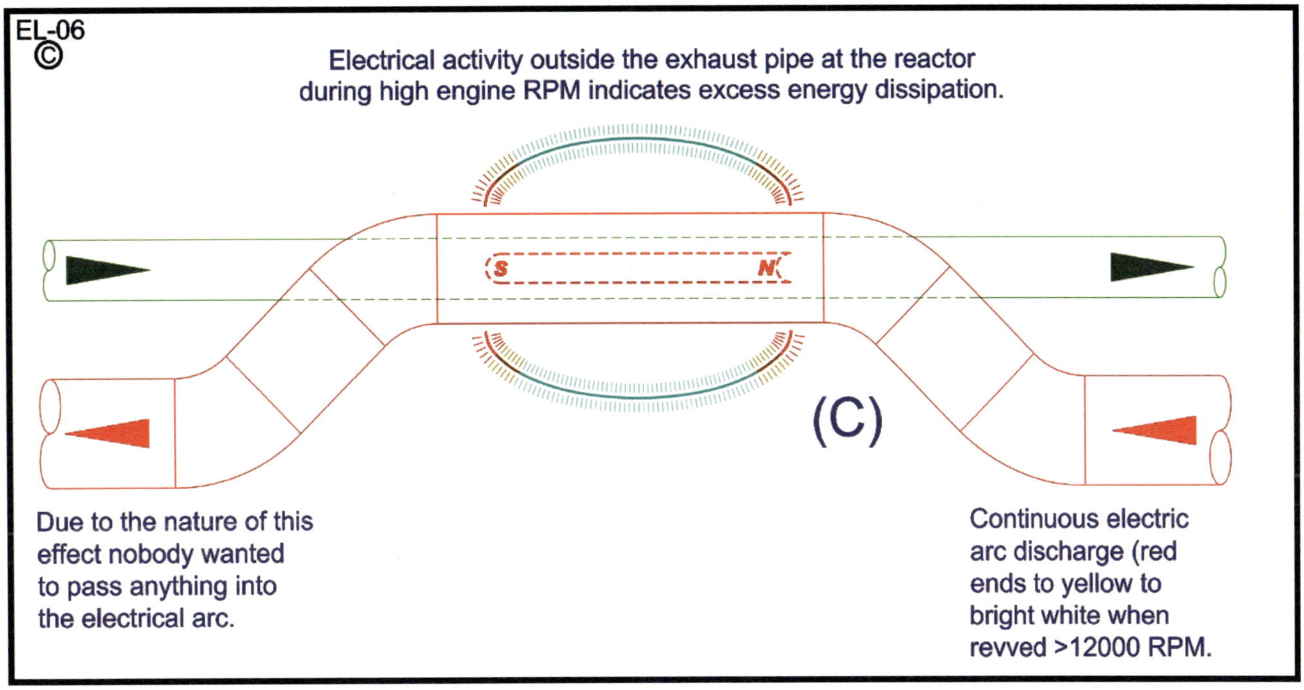

During another test, or should I say demonstration with about 15 witnesses, we revved the engine up and the entire exhaust pipe reactor rod and reaction chamber appeared to be gone. These events seemed to replicate the Philadelphia Experiment. At the time, although scary, I was very surprised to see all these football fanatics running at high speed to get away from the reactor and hide behind walls, trucks, and trashcans.

When I explained this phenomenon to several scientists, seeking answers as to what could cause this event, all of them recommended I not do this test again without doing it remotely from at least seven miles away. Preferably, I should do it in the middle of the Nevada desert!

In December 2000, I received an email asking me to check out an article written by a PhD from A&M University. The article was entitled the "White Hole Theory." As I began reading the article, I realized the person who wrote this believed it would be the other end of a black hole. As I continued reading his thesis, it read more like a description of what GEET is and how it works from the vortex at extremely high speeds to pull electrical energy into itself. At the north end, there is a ball of white light called plasma. I could not describe the GEET any more efficiently than this professor could, who had never seen it.

MA-05
©

White Hole?

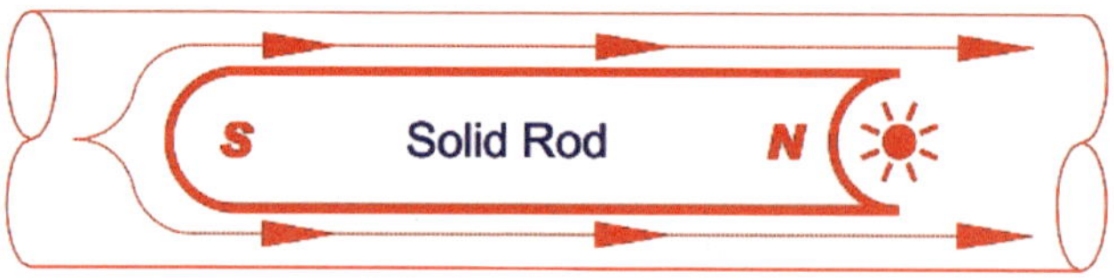

A few people have built a GEET reactor so the north end would be a glass port so they can look down the length of the reactor at the plasma field as it generates. Most are getting a blue light, which is only a partial plasma field. When a full plasma field is being generated, a bright white light radiates through the metal and is visible from around the room.

A common question asked over the years is "Does the engine lose horsepower?" This is a little bit more complicated than a simple yes or no. The answer would depend on the fuel and your settings, the proper size of the rod, and the amount of heat exchanged to cause a reaction. We had 10 horsepower engines actually put out 28 horsepower, but the same engine on different fuels might only put out 8 horsepower. We have found ways we can get more energy than the engine design should be able to output.

Just because you can get three times or more horsepower, it is not recommended. The transmissions and drive components cannot handle such power. You could cause many problems trying to use the extra power. Do use caution when trying to do this.

Some tests completed in France tested the metal of the rod after it had been used. It indicated an intense amount of heat applied to the rod at the halfway point. The hardness of the metal at this point showed the equivalent of over 400°C spiking temperature within just a few millimeters. Please keep in mind this 400°C jump in temperature happened in a vacuum.

You should also keep in mind that water will boil at 40° F in a vacuum. During a good run of GEET, a bubbler will have ice on the outside of the chamber. If you add sugar or salt to the fuel being vaporized, creating a longer molecular chain, you can expect to see ice forming on the exhaust pipe and muffler even if the air temperature is 110°F.

This makes it very difficult to run a GEET when the outside air temperature is below freezing. We have had ice form in gasoline and caused it to solidify when ambient temperatures are at 40°F to 50°F. In these models, we do not use water as a fuel. It was very embarrassing during our tour. I forgot we would have no heat coming from the engine, and therefore no heat or defroster in the suburban. We had to install two electric heaters just so we could get out of the state of Michigan without freezing. So getting good mileage may sound good, but you have to look at all other aspects associated with getting good mileage. On tour, I used my pocketknife to poke a hole in the upper radiator hose. Travelling 80,000 miles with a hole in the radiator hose, without losing any antifreeze, validates the coldness of the engine when retrofitted with the GEET.

When running our first converted Waukesha 240 KW generator, we disconnected the radiator completely and just circulated water within the engine. During our test runs of three and four hours each, we could never get the temperature of the engine up more than 1° from ambient air temperature. Some of the benefits of this would include longer run times, never needing to change the oil, not needing a cooling system and no pollution.

Two young boys, 10 and 12 years old, from Palm Desert, California, used GEET as a science project and won first prize. Although the teacher would not allow them to bring gasoline into the building because of state laws, the engine still ran great when they pulled the rope. They were the most surprised, since all they had put in the fuel tank was Windex and pickle juice. I had instructed them, as I do everyone, to use at least 20 percent gasoline to stay politically correct.

The Truth Will Come Out

In the early 1990s, a reporter from the Marin Independent Journal went to a local gas station to interview me. The reporter got back to me, was very rude, and said I was a con man since Briggs & Stratton claimed I was never at their facility doing any testing. When his editor received a copy of the visitor's badge showing I was there, he called Briggs & Stratton to find out why they told the reporter they never heard of a Paul Pantone or GEET.

The people at Briggs & Stratton Research told the editor again they did not have any knowledge of GEET or Paul Pantone. He asked them for a fax number and asked them to hold. He faxed them a copy of my badge. When he came back to his phone and asked if they received a copy of the badge, they were upset because I had taken the badge off their property. This indirect admission of my being there was all that the newspaper needed, and they printed the story on the front page.

Over the last 37 years, many people who have seen the GEET will not admit they even saw it. Some scientists who had written letters of testimony to us contacted me and asked me to remove their testimony from my website since it was causing them problems.

July 3, 1995

Andreas Kurt Richter, Ph.D., P.E.
6814 - 278 th. St. N.E.
Arlington, WA 98223
Tel: 360 435 8310
Fax: 360 435 2155

To whom it may concern :

It is my opinion that Mr. Paul W. Pantone has invented an amazing energy device or engine with potential as yet unheard of.

Recently Paul W. Pantone and his associate [redacted] took such an energy device apart while I watched . They let me inspect every single part of it and I could find no evidence of trickery . Then they put it back together and ran it with a mixture of mineral oils and water for fuel .

I witnessed the engine driving a mechanical load built to my specifications, operating with unheard of efficiency and a surprising lack of pollution in its exhaust . It was also run long enough to prove that it operated well past the point of just using up the more volatile components of the fuel mixture .

I am a consultant to Paul W. Pantone in the search for the scientific and technical explanations to understand the operation of this energy device . According to my present knowledge it should not work and I would not believe it, had I not seen it with my own eyes . The available literature too has not yet provided me with any answers .

Sincerely

Back in 2005, there were a few hundred people in the U.S. that placed GEET on their equipment. After the court and other agencies harassed them, they all wanted to be removed from our files. About fifteen of them appeared at my hearings in Utah to show the court that GEET did in fact work. This made Judge Royal Hansen very mad, and he asked the bailiff to get all of their names and addresses. He refused to look at even one engine.

Reactor Configurations

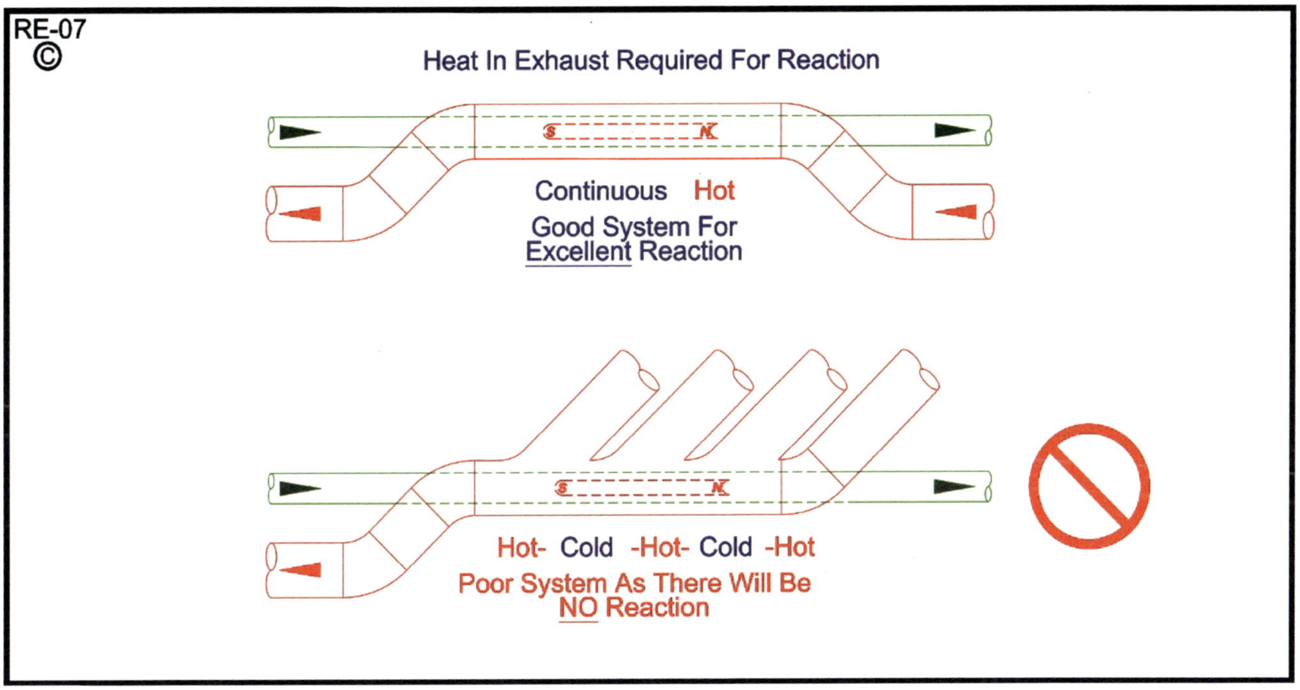

Before you fully validate something, you must try several different ways to see if you can improve on the idea, or if your final idea is the best. We tried three, four, five, and even twenty reactors at the same time coming off one exhaust pipe. We tried placing one reactor after another, and they all failed. One of our failures was to drill holes through the existing exhaust manifold of an engine so that the cylinders came from the block of the engine into a common rail. Through these holes we would insert the reaction chamber through the common rail proper. The result stopped any reaction from occurring. To get the best quality of heat coming from the engine or heat source, you need a smooth continuous flow of heat leaving the engine after all the exhausts have merged into a single pipe.

Some people wished to simply mount the reactor pipe by hose clamps or spot welding it to the side of the exhaust pipe to cause a heat exchange. This process only affected part of the rod and it made a vaporized fuel, not plasma. Others tried shaping an exterior additional piece with a tube through it, but this also failed miserably.

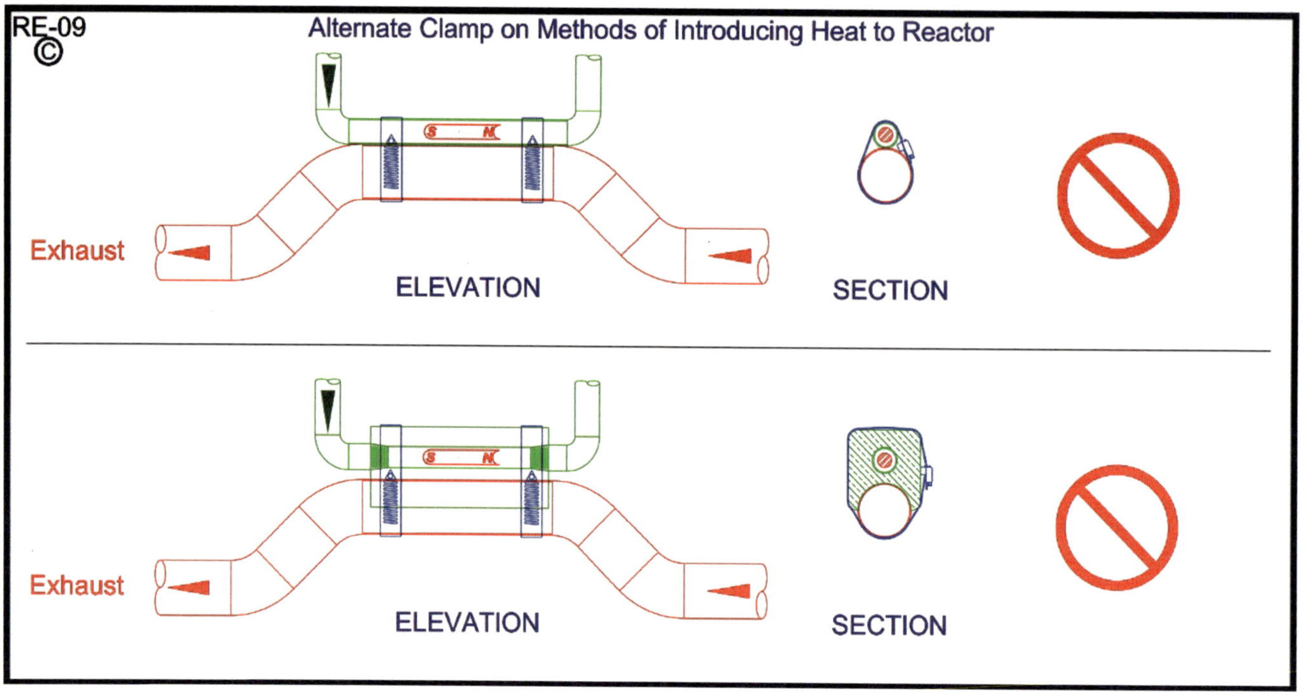

If you use a loop in a fuel line in a horizontal direction around the exhaust pipe, it will run fine as a vaporizer and give you about a 30 percent increase in mileage. However, this would violate the wet fuel near an exhaust pipe law put in the Agricultural Bill. The problem with this method comes when you shut your engine off, the vapor condensation will stick to the pipe and settle to the bottom of the loops and some of the particles will harden. After a period of time, these hardened areas will completely block off the loops so that the vapor can no longer get through. Another attempt to get better mileage and save money was to use rubber or plastic lines with the vapor coming in and out of the reactor. Yes, this will work, but not for long.

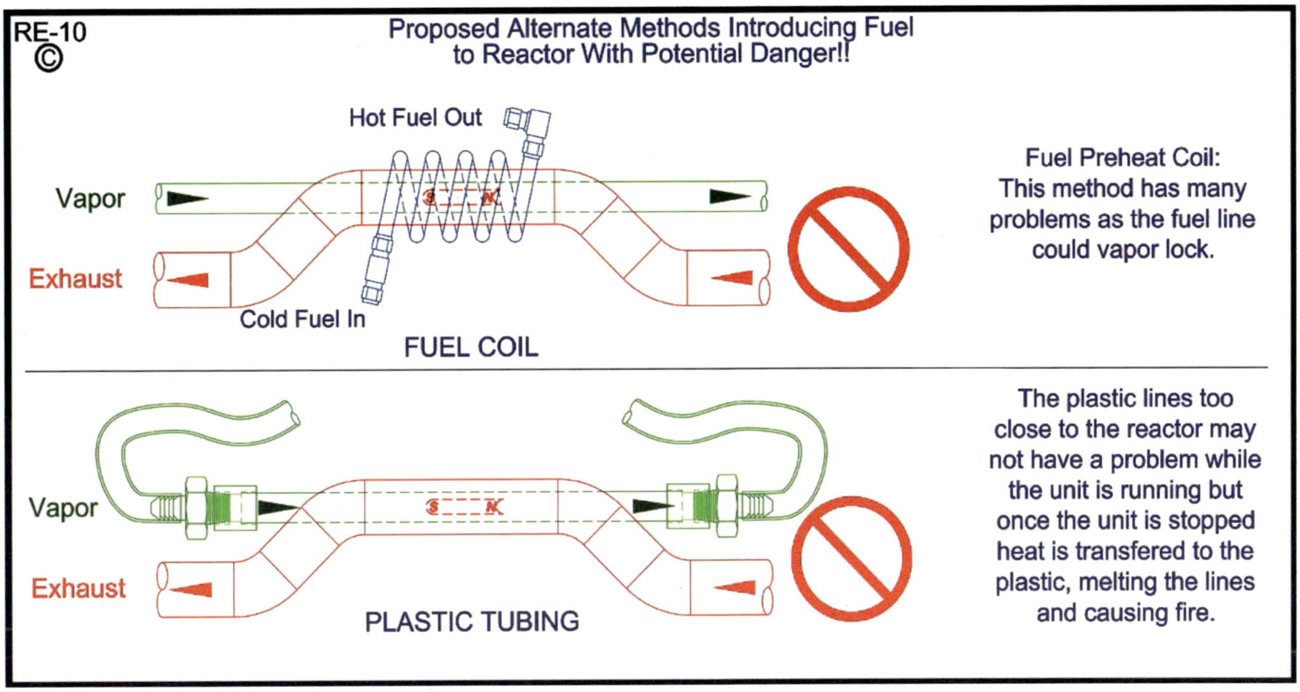

After running it for a while and then turning it off, there will be a certain amount of heat transmitted outward from the metal until it hits plastic or rubber and melts it. Then you will find you have a vacuum leak and the engine will not run. To stop this from happening, we have told people to use a minimum of six to eight inches of copper and keep it a distance from the reactor. This will allow the heat not to affect the plastic or rubber.

There have been many times when monitoring the engine the heat within the reaction chamber far exceeds the heat from the exhaust. Yet the final exhaust coming out of the reactor is sometimes as low as 30°F on a 110°F day. The first reaction occurs in the center of the rod where the largest spike in temperature is located. If the engine is decelerating quickly, the reaction pulls heat from all directions through the metal. You can see this with the digital handheld thermometer. If the temperature coming out the exhaust begins to drop while the engine is running, then the plasma reaction is working properly.

There are numerous ways to increase the volume of the fuel, whether by using air like exhaust or hot water from a radiator system. If you use gasoline or vapor and feed it through the hot air or liquid in the opposite direction, it causes an expansion of the fuel and some heat transfer which gives you about a 30 percent increase in mileage. However, this opened up a completely new Pandora's box. A simple heat exchanger within a metal box using hot water for the radiator did give people the 30 percent increase in mileage, and a few people even got double their mileage. A few complaints came back from the people doing our testing at different latitudes and elevations. It seemed the mileage went down when the air temperature was above 72°. Performance also went down due to atmospheric changes.

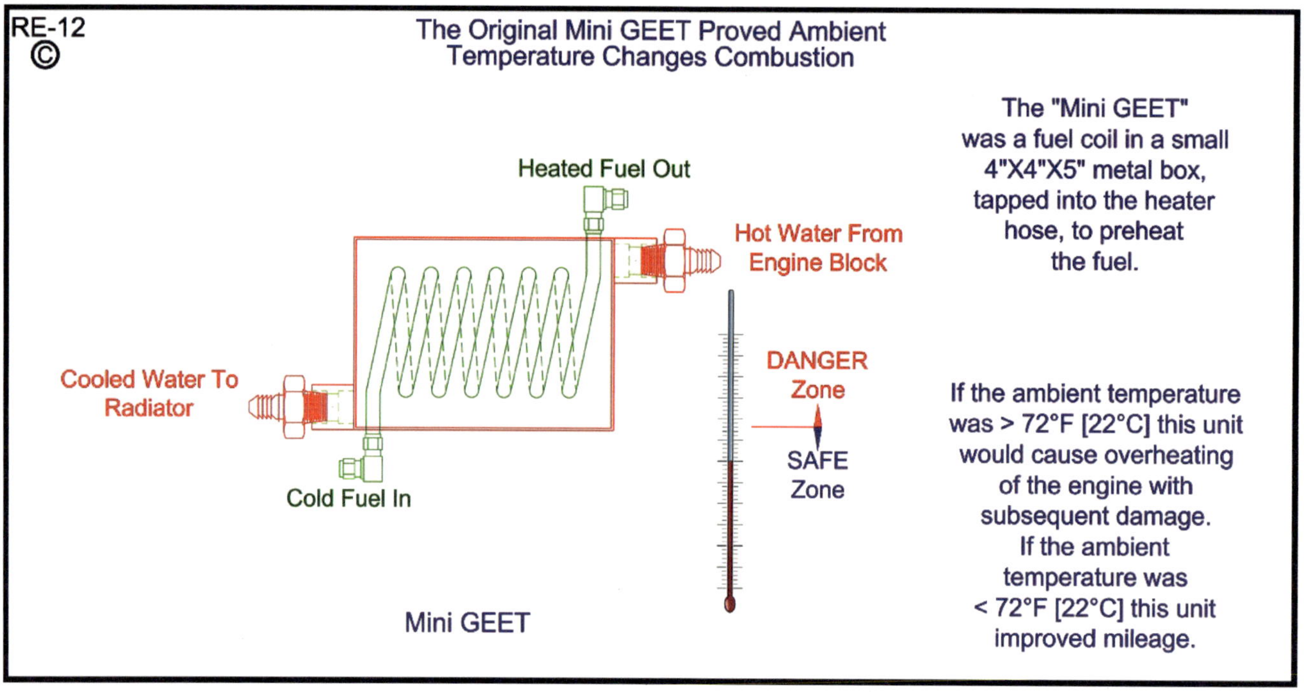

This led us to an entirely new discovery and invention, or maybe a whole series of inventions. We have not found the exact temperature to blend the air and fuel to give an efficient burn for each engine. Based on the test so far, there appears to be a point where the temperature is very critical and gives you optimum performance. Every new car made today with the proper air temperature control and fuel temperature could give you a 30 to 40 percent increase in mileage, or maybe even more.

GEET Gas Is Not The Same Old Stuff

Prior to my going on the three-and-a-half-year 80,000-mile tour, my brother-in-law demanded to tune up our Suburban, since he was a trained "Mr. Goodwrench." After he hooked everything up and marked the flywheel, he started the engine. He was looking at the flywheel and turning the distributor. Nothing changed when he did this. The engine ran smoothly, and it did not respond to the position of the distributor.

He then turned off the key and the engine continued to run very smoothly. He got under the hood again and began pulling off all of the spark plug wires, while the engine continued to run effortlessly. At this point, my brother-in-law came to get me. I held a straight face as he explained the engine would not shut off. I offered to help him and I went outside with him to the car. I looked under the hood first and saw that he removed the spark plug wires.

Then I told him the engine sounded pretty good. He explained to me that it should not be running with wires pulled and the key off. I went into the cab area and turned the ignition on for just a second. Then, as I turned the key off, I slid a valve hidden under the dash to open air to the intake of the engine.

This broke the vacuum and GEET stopped working. I got out of the cab of the truck, came back to the front, and asked him if there were any other problems. He shook his head no and proceeded to remove his tools from the engine compartment.

I asked him if he was done and he snapped at me "yes!" I got back into the driver's seat as he was telling me it would not start without the spark plug wires hooked up. As he was saying it, I closed the secret valve, turned the key, and the engine started right up. With that, he threw his tools in the back of his car and slammed the trunk. That act wrecked his toolbox and his trunk as he forgot to close the lid on the toolbox. As he drove off, I hooked up the eight wires he removed.

When a GEET is running properly and efficiently, it does not need a spark. This has only happened to a few pieces of equipment where we installed GEET. However, it is always entertaining to see the look on mechanics faces. Without thinking about it, we drove to Denver, Colorado for a conference. While there, we had the hood of the Suburban opened with the engine running. A mechanic came up to me and asked why the distributor was sitting there moving back and forth. I glanced over and I realized my brother-in-law had forgotten to tighten it down. Since I did not have any tools, I asked the mechanic if he might have the right size wrench to tighten the distributor.

Power And Cleanup

For many years, while trying to track down all the inventors I could, as well as race car drivers, I came across many of their secrets. One racer told me of a disqualification because he put a mothball in his tank. At first I thought it sounded rather funny, but then I realized a mothball is 99 percent naphthalene. I decided to try it in my fuel, and lo and behold, it worked. It definitely gave me an increase in horsepower, but quickly increased the running temperature of the engine. I was not aware it could be against the law.

I would never recommend using mothballs with any internal combustion engine as it does cause excessive heat and will burn the valves. I am sure it is not healthy for the pistons either. A race car driver who races Porsche engines had told me he put mothballs in his tank prior to taking a run on the racetrack. His car had an octane reader and it went from 89 to 180 when the naphthalene hit the sensor, and did he have power! He said it was not quite as fast as running nitrous, but it sure boosted his power and made it "super-duper" gas. Nevertheless, since talking about this on a radio show back in 1985, we now see that mothballs cannot be sold in the state of California. It says so right on the box. Also on the side panel of the box is a note that it is against federal law to use mothballs for any other purpose than killing moths and their larvae.

When building the GEET and trying to determine the specifications, first do a compression check on your engine cylinders. A Mazda donated to us for our school was found to have compression of the four cylinders at 30, 30, 120, and 30. This would indicate a buildup of carbon holding the piston rings closed or keeping the valves open. After running the engine for about four hours using GEET gas, compression was checked again and all four cylinders were at 120 pounds. We had to cut the exhaust pipe off the vehicle and point it straight down so the debris collected in the exhaust pipe from the engine would not plug up the catalytic converter and muffler. At the end of the exhaust pipe was a pile of tarry black substance that was difficult to clean off the concrete. We reattached the muffler and catalytic converter and tested the emissions. They would have passed at any smog certification station in any city. Unfortunately, the reason the car was a gift was it used about one quart of oil per hour.

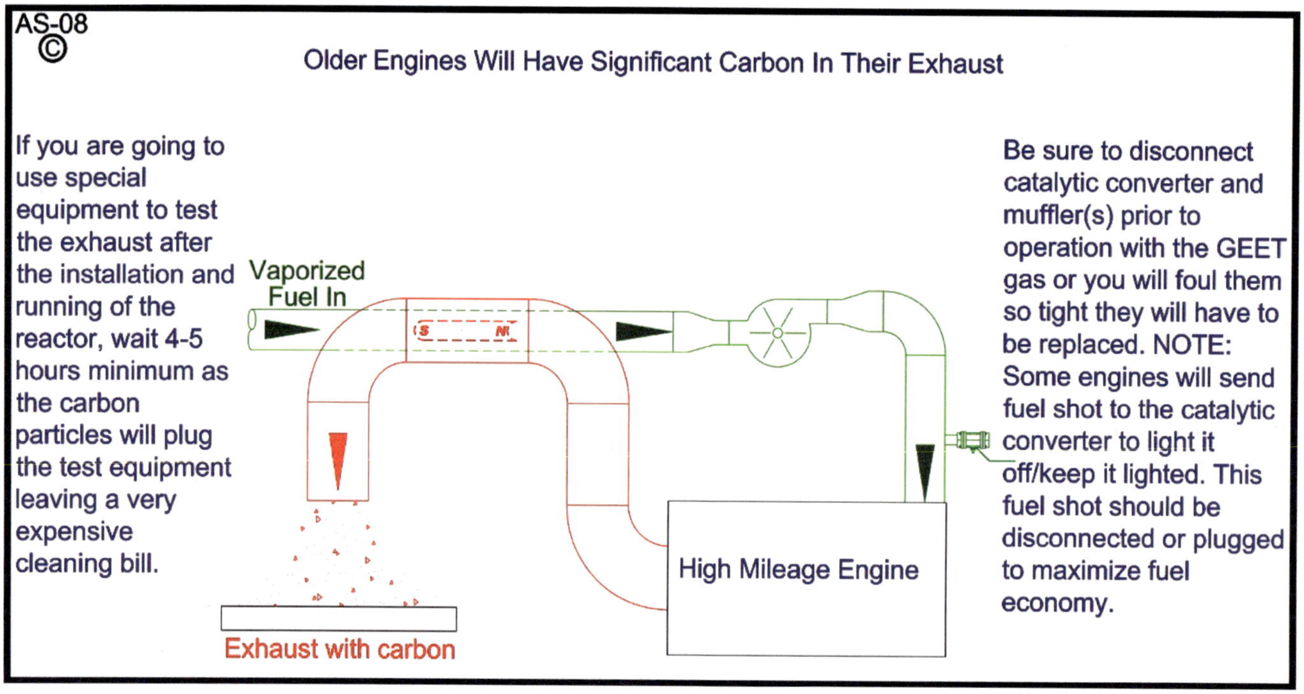

We were definitely very lucky with that car since the rings were not broken and the walls of the cylinders had not been seriously scarred. We were able to save the engine without even cleaning off the outside of it. Mechanics who looked at the engine wanted to know how we ever got the engine completely overhauled without removing any of the dirt and oil sludge. Since I do have a sense of humor, I told them we put little men down each spark plug hole into the cylinders, gave them each a little brush and said we would not let them out until they finished cleaning the engine properly. With a straight face, I said these are the same guys that are related to the scrubbing bubbles. The mechanics who saw this engine before we began working on it smiled and realized we knew something about engines they did not.

We decided to have some fun. We made the carburetor intakes to fit a 350 Chevy out of glass which we glued together, another out of a beautiful cherry wood box, and another of steel. The one I liked the most to get a smile out of mechanics would have to be the wooden box. In 2001, when I opened up a school in West Valley, Utah, I went to a cabinet shop across the street from the school and handed the owner a gasket that goes under the carburetor of a four-barrel carburetor. I asked him how much it would cost to build such a box with the bolt pattern like that on the bottom. He smiled, looked around behind me and out the windows, and said it would be $40 and he would have it done in an hour.

Two hours later, I went back to the cabinet shop and picked up my new carburetor. It was just a wooden box about three inches squared, mounted on top of the platter piece which would then bolt to the top of the engine. I had students in class that week who volunteered to put the intake on top of our engine. After a little fine-tuning, I drove the truck across the street with no hood and asked if he would like to see a wooden carburetor. The owner smiled and leaned way back in his chair, putting both hands behind his head and locking the fingers, and said, "Do I look that dumb? Do you think I'm going to fall for this?"

His wife was there. She went over to the window, looked out, and could see that the wooden carburetor was on top of the engine. She invited her husband and the guys working in the cabinet shop to take a look at it. The guys in the back shop came out to look and they were all impressed, but the owner still refused to move. He said he knew this had to be Candid Camera or some kind of video thing and he was not going to fall for such a ridiculous idea as a wooden carburetor. To this day, I do not think anyone could convince him we really ran that engine with a wooden intake. If you think this technology cannot work, then you should not be reading this book. You do have the right to believe anything you want.

The next time you buy something that looks complex, try to imagine what it was like before it got that way. As with all things we have today, something does not exist if someone did not take careful thought and build it. We have too many things today, which have limited us in our ability to think. If everyone believed there is only one way to do something, then we could never have new discoveries and a better way of life.

Small Rod, Big Results

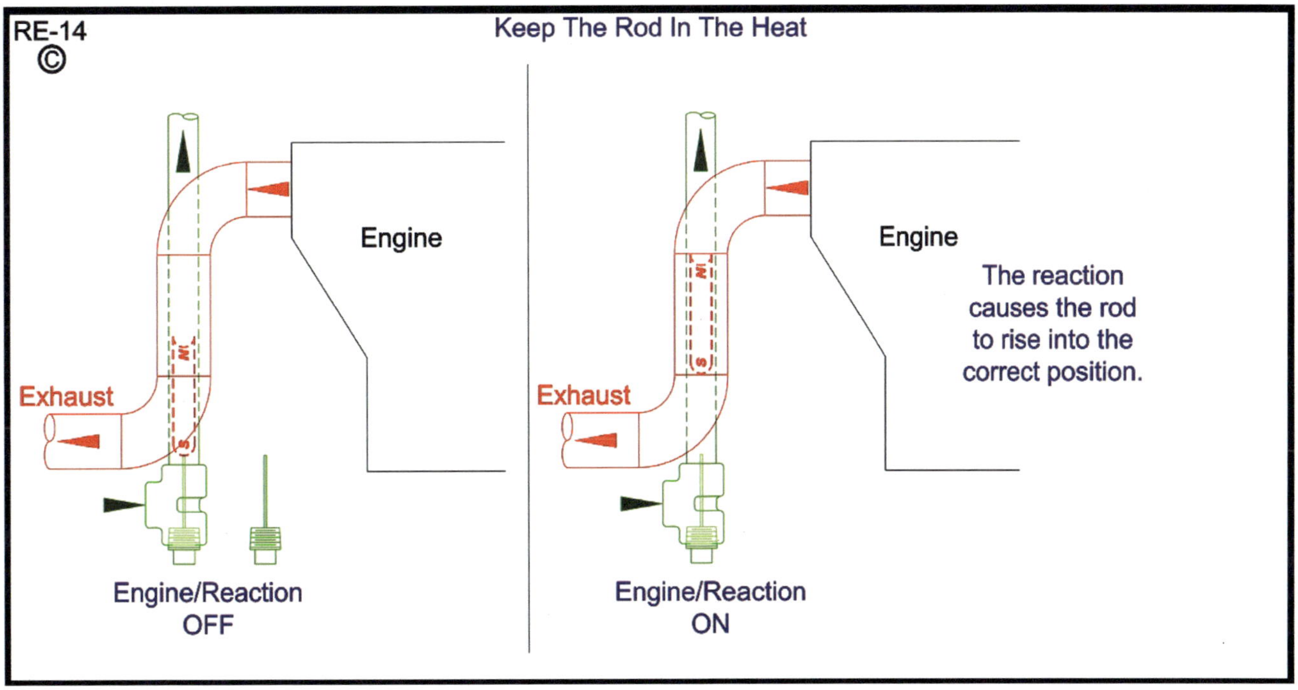

As shown in picture number RE-14, we have on the left side a threaded pipe plug with a small rod sticking out of it to hold the rod in place prior to running. When you remove this plug, it will allow you to replace the rod or inspect the rod without disturbing the rest of the GEET application. It is very important, should you have any problem with the new fuel you always check your rod first. If the fields do not flip, and north is always up, you have a problem. If north and south are not at the north and south end of the rod, but above it or below it, you have the wrong size rod for the fuel you are using. A simple compass will always tell you how long it needs to be.

Not In America

In 1994, Brigham Young University invited me to test the GEET at the nuclear laboratory. I went for three days of testing and was very happy with the results. It was on the third day that one of the scientists turned to me and said, "Transmutation of elements." He said this was just as exciting as when they found out cold fusion really worked and they did it right there. As they were saying this, Dr. Jones turned towards a glassed in box with hand gloves and said they did the test on cold fusion right there.

I was rather perplexed since I had heard BYU's position was cold fusion did not work. The good doctor told me the man who writes his paycheck told him to take the position that it did not work. Not wanting to lose his job, he did what they asked. He then turned back towards me and said he would never publicly admit that he saw the GEET work either.

I was getting used to this. People who were sent to work with me from other universities would suddenly come back, grab all of their things from my shops, and say, "I'm sorry but I cannot work with you and I can't tell you why."

This horror story of knowledge coming forward will make even the strongest person feel sick. As I spent three and a half years on the road, we were not welcome at any public venue. Even GM, Ford, and other car companies wanted to see the technology in action. However, they did not want to see it on their property. They kept asking us to meet in private locations or private shops away from their corporate property.

From Lawrence Livermore National Laboratory, Sandia National Laboratories, and White Sands Missile Range, to all of the other government assisted laboratories, I had instructions to never bring a unit for them to test. A small handful of these people told me if I ever brought a GEET to their facility, I might never be seen again because the technology would never leave their property. A few of these nice scientists warned me back in the early 1980s to be very careful what I said on the phone, as all calls are monitored. They said when I had technical questions I should call and ask them using hypothetical terms, indicating I never really built one, as they were allowed to discuss technology in theory. They did not want to admit knowing that GEET really worked. Note: If you are an inventor, reread the above again and take it to heart.

The part that hurt me the worst was being told not to do this in America. My goal from the beginning was to create something that Americans could build in America. The profits would go back into the economy to help this nation and the people who live here. I realize now that it was just a dream trying to give this country the boost it needed to be back on top. It would have been a lifetime achievement. I heard this from presidents, vice presidents, and CEOs from almost every American company I contacted. They said the same thing: take it to a third world country. America is quickly becoming one. Sadly, maybe that is how I will get my dream right here in America.

Totally New Fuel

At BYU, we ran an engine with GEET, hooked it to a bubbler, and used crude oil as a fuel. I will never forget this experience. I saw the scientists getting all excited as their mass spectrometer indicated the parts per million or parts per billion of elements from the bubbler to be 72 different elements. After the vapor went through the reactor, the scientists extracted the vapor and fed to the same mass spectrometer. It now showed only 12 or 13 elements coming out. What got the scientists excited was that there was no hydrocarbon, sulfur, or heavy elements. There was an abundance of hydrogen, helium, and something between the two, with an atomic weight of 2.4.

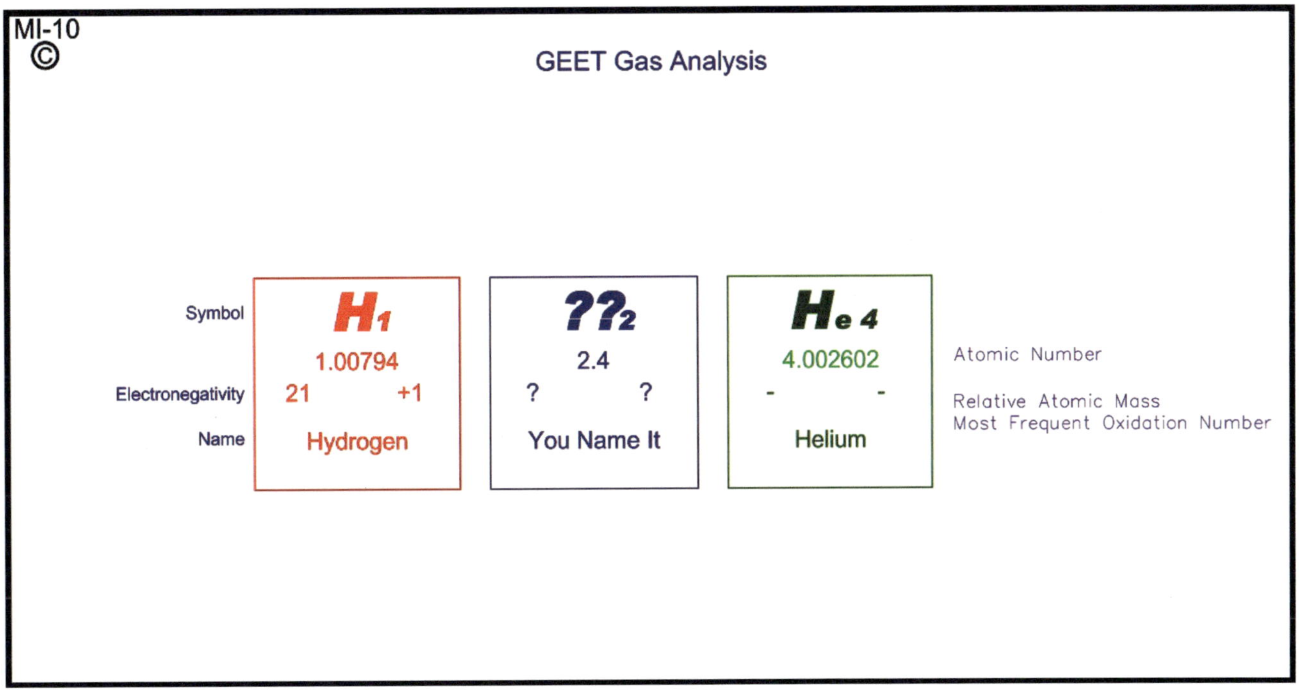

Could there be a new element? Yes! According to my dreams, there would be five more elements found over the next ten years. This is if my dreams are accurate. I am just a man, and I put my pants on one leg at a time. Although I dreamed of how to build the first GEET when I was about eight years old, it took until I was 33 to do it. Then again, when I was five years old, I did dream of how to build a magnetic motor when I was 34. There are things I dreamed of when I was 5 to 6 years old that I have not done yet.

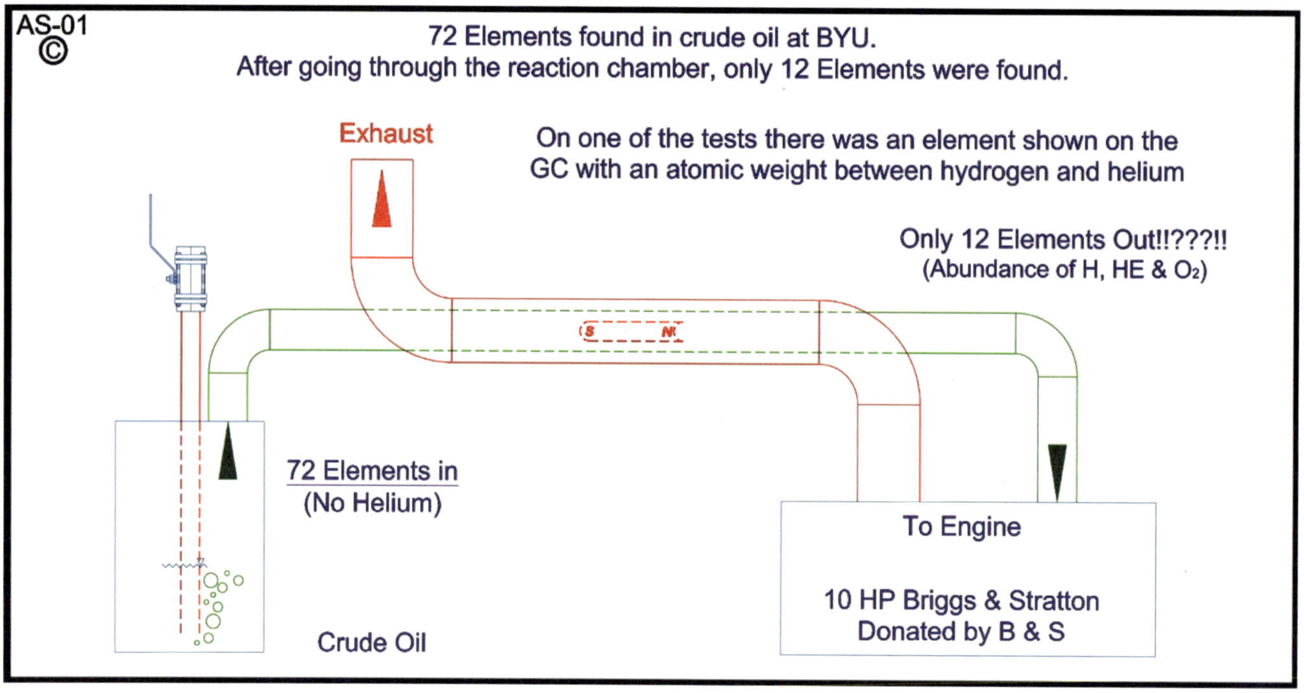

Back to BYU. Once the new fuel went through the engine and came out in the final exhaust, the quality of air was almost identical to the air in the laboratory. If you ever started up a lawn mower in a closed garage, you will know how much pollution they normally can put out. Normally, I can only take about 10 minutes of the garage door closed and my eyes are burning. During one of our tests, my technician ran a 10 hp engine on crude oil in a 10' x 12' room with all the doors and windows closed for about eight hours. The quality of their air was cleaner than the air in the next room.

For people who really want more power, this is easy to achieve with GEET. So far, we have not had the guts to push the upper limit, but it should be done in a room by remote control so that no one will get hurt if the engine blows up. Since we have achieved over 12,000 RPM from engines that are rated for 3,600 RPM maximum speed and not blown the engine up, we don't yet know the upper RPM limit.

During a presentation, Dr. Wood was standing about four feet away from a 350 Chevy engine which was running and hooked to a generator. He turned and asked me to please start the engine. I looked over at the gages and they indicated the RPM was at about 70 and water temperature was normal room temperature, as was the oil pressure.

I leaned a little to the right and could see the fan blade was turning nice and slow. I pointed at it and explained out the engine was running. Dr. Wood turned and watched the fan blade moving very slowly; you could see how excited he was getting as he turned towards me and said that 350 engines could not idle at that speed. I showed him the RPM gauge and he was still insisting the engine could not be running. Not only was it running, but it was also running very smoothly. The blend of the fuel was so perfect that the burning occurred in the dead center of the piston with each stroke of the engine.

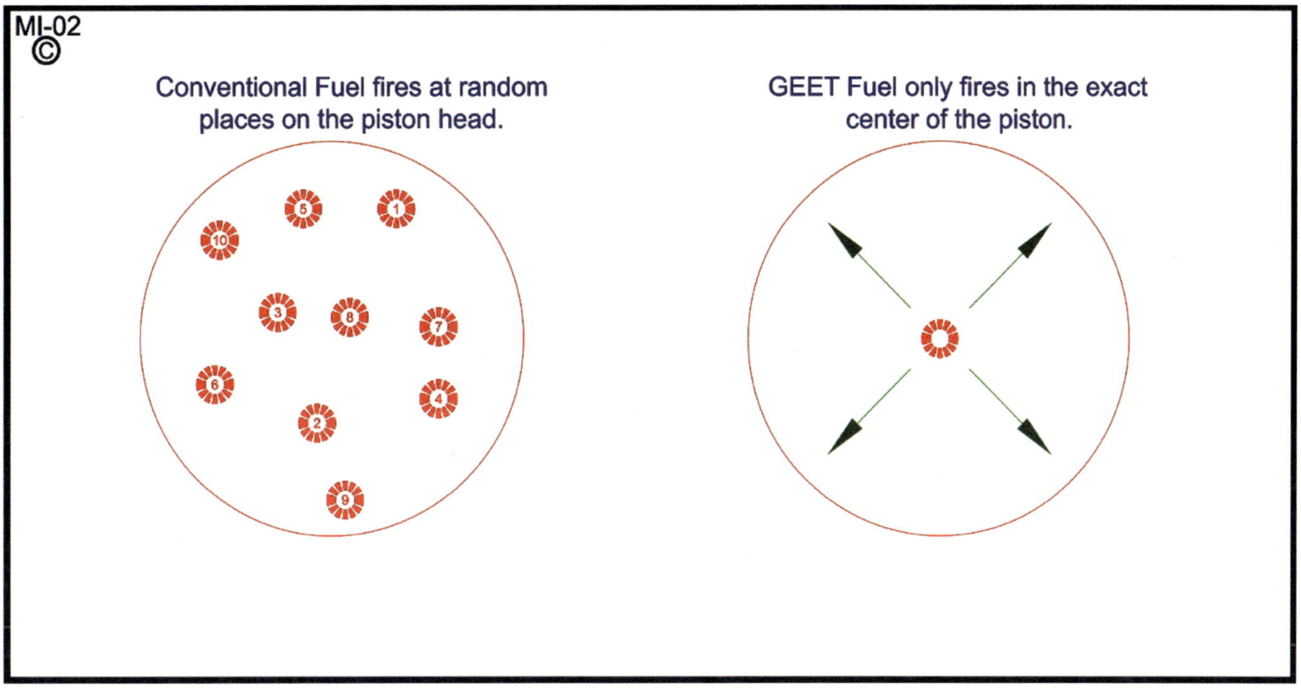

More Power

Some people want all the power they can get, and they simply ignore our warnings. These warnings are there for a reason. It is safety for you and your equipment. Equipment based on the first three states of matter is designed to run between a low speed and a high speed deemed to be reasonable and safe for that specific piece of equipment.

I helped install a system for one student on his monster truck. By adding a pump between the reactor and the intake of the engine, I could pull the vapor through the reactor and pump the fuel into the intake of the engine. You should only do this with a pulley driven device. Your foot or a throttle cable controls it to regulate the speed of the engine so the vacuum pump operates safely.

Once the unit was operational, he removed the vehicle from my property and took it to a racetrack. I am sure glad no one was in the bleachers when he turned on his electric pump to the intake of the engine. The driveline got ripped out of the vehicle and thrown sideways over the bleachers. If the driveline had hit a crowded bleacher full of people, it could have killed many of them. Not to mention the damage done to his truck where his two-piece driveline ripped a hole in the floorboard as it exited the underside of his vehicle. In total, he only lost about $23,000 that day, but he learned a valuable lesson. When all else fails, follow the instructions.

Not all pumps can withstand the temperatures produced by GEET gas. Alcohol powered racing vehicles operate with such a pump and are capable of handling such high temperatures of vapor. If a pulley controls the pump, then the speed of vapor delivered is controllable by throttle connections and become safe to operate. You may still wish to put a governor in place to control any chance of over speeding the engine.

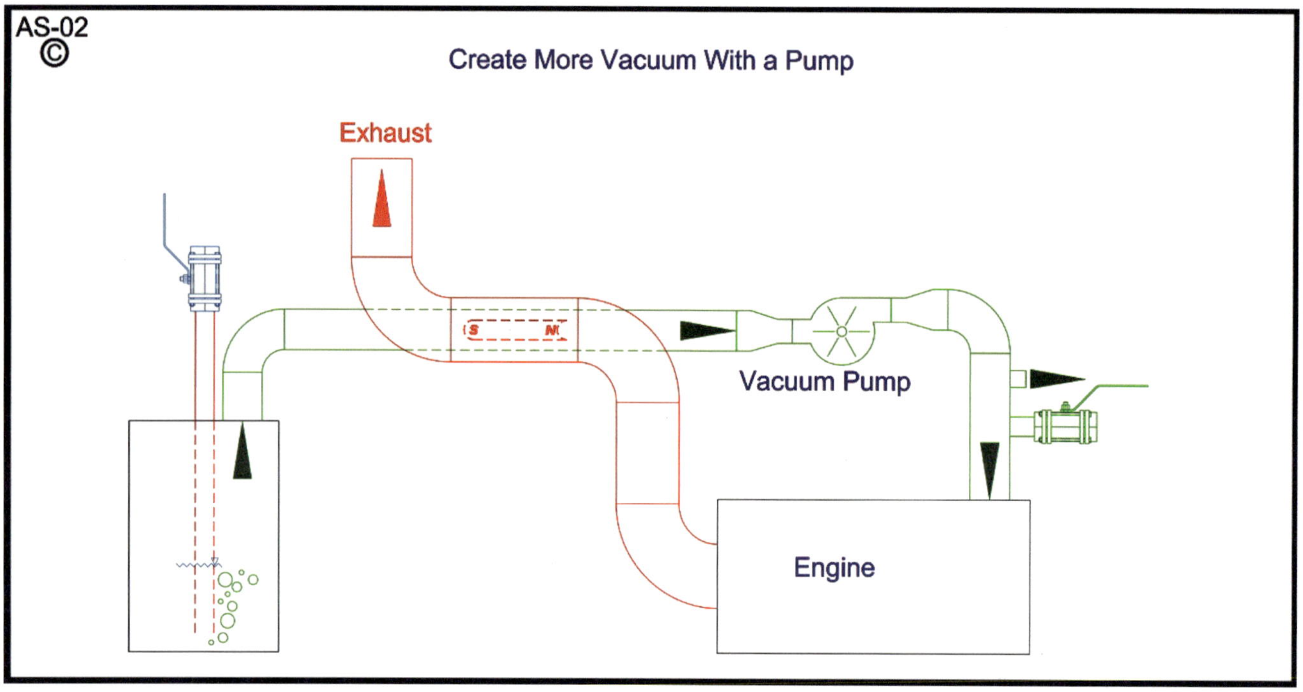

One company from the Los Angeles area sent a man to come to school to find out how to modify racing cams using the GEET technology for fuel delivery. When this mechanic found out the timing had nothing to do with the running of the engine and only required the piston be moving up and down and you had the proper air fuel mixture settings to operate, he realized any cam could be used for racing. Now in case you did not catch what I just said and understand it, let me explain.

Normally, a four-cycle engine operates with one stroke downward pulling air and fuel into the cylinder, using power from the engine to operate the stroke. Another stroke pushes up the piston to compress the air fuel mixture, causing a loss of power to the engine. The power stroke is where a spark ignites vapor within the cylinder and the explosion pushes the piston down and causes energy to crank the engine. Lastly, there is the exhaust stroke, requiring power from the engine to push the exhaust out. This means only one of the four cycles is producing power and three of the cycles require power from the engine.

With GEET, the strokes are much different. First, a pump pushes air and fuel into the cylinder, giving power to the engine. Then the compression implosion stroke occurs where the piston is rising and the compression causes GEET gas to implode, pulling the piston to the top. This becomes another power stroke. The power stroke is self-detonating as the piston begins to go down, causing more power to the engine. The final stroke happens because of the restriction of the reactor rod assembly within the exhaust pipe. It causes a vacuum on the exhaust of the engine and is another power stroke for the engine. So now, you have four out of four strokes creating power.

The implosion pulls heat from the walls of the cylinders causing a cooling effect on the block which is equal to the heating effect of the detonation. This system never allows the engine to heat up enough for the hot water thermostat on the cooling system to even open. This is why I was able to drive over 80,000 miles with a hole in my upper radiator hose and it never leaked.

There Are More Transmutations

Many of our experiments led to discoveries and situations we could not find referenced in any science book. One such test appeared to turn gasoline into paraffin. This amazing result happened when there was a slight modification to the location of the bubbler due to a lack of space when the GEET powered engine was hooked to a compressor. Placing the bubbler within a few inches of the reactor causes a visible change to the substance in the bubbler.

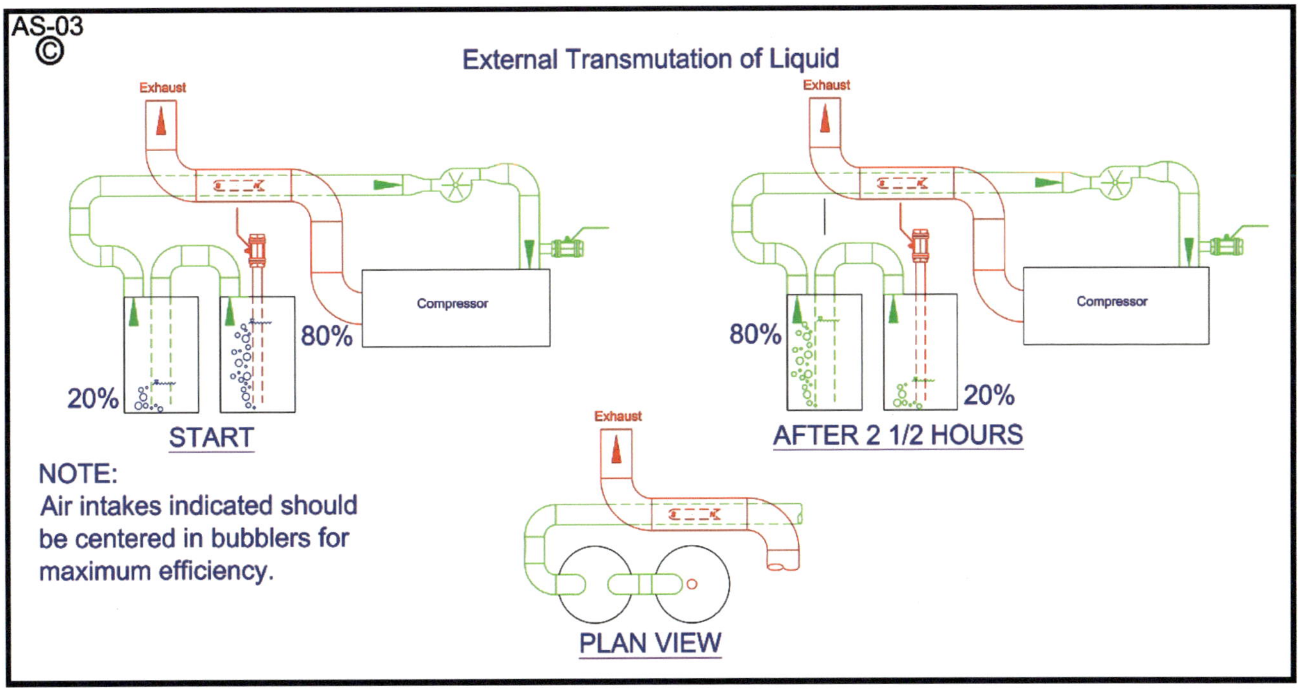

When you place the fuel within six to eight inches of the reactor while the reactor is running, something is happening to the fuel to make it become denser. During this experiment, the gasoline turned to a white gummy substance resembling paraffin. We put cardboard into this paraffin type material and then ignited the end of it with a match. The cardboard burned like a candle, very slowly. We will do a great deal of testing in this field to see what qualities of paraffin we can produce.

Different Configurations

The first servant engine we made back in 1984 consisted of a pump system and control valves. This was so we could pull the fuel through the reactor. Part of the fuel was used to run the engine that was creating the fuel. There was a T in the line to allow a second engine with a control valve to supply the GEET gas to the engine. There were no modifications to the second engine except the air intake. During a series of tests we did over the next two years, we found the heat from a 3 ½ HP engine could produce enough fuel for a 327 cubic inch engine.

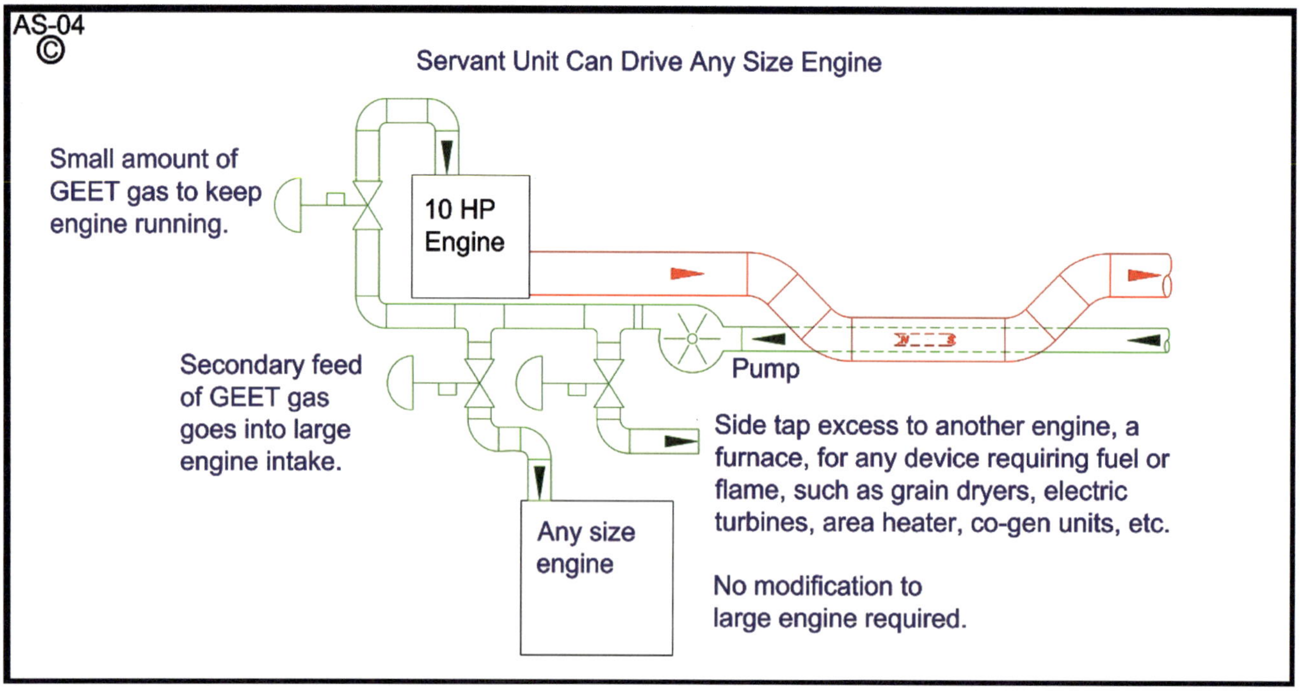

It took about 12 years before I realized you simply had to split the exhaust pipe and have two reactors. One reactor fed the servant engine and the other fed the new engine. When using this method, we scaled up to a 10 HP engine, and ran a second engine which was 440 cubic inches.

This same twin GEET pipe can be used on just one engine with two separate incoming fuels.

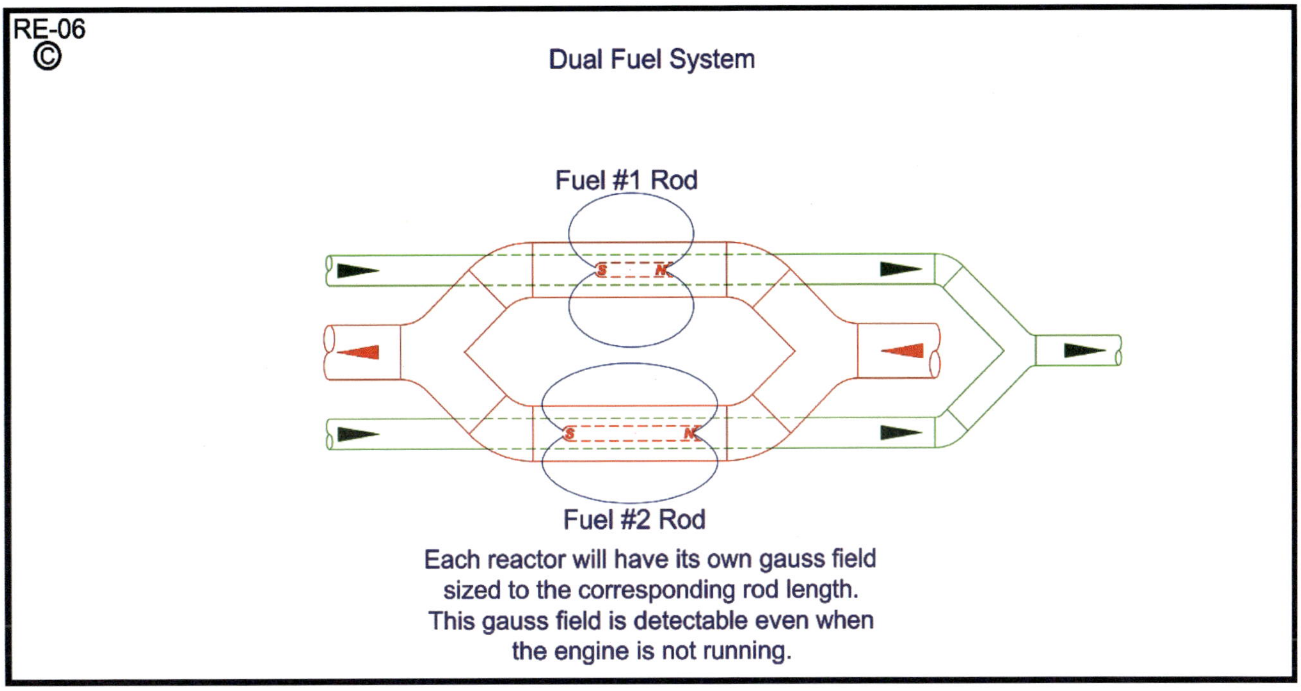

With all the engines, we had to have a vacuum break valve or vacuum relief valve within three to seven inches of the intake of the engine to stop the GEET vacuum and enable it to shut off. Just turning off the electrical system will not shut off the GEET system if you are getting a full plasma fuel.

One of the most unusual discoveries was that a gasoline engine runs better on diesel than gas. Moreover, a diesel engine runs cleaner and with more power with a gasoline and water blend than with diesel. Who knows, you may discover other things far beyond that if you try. Just do so at your own risk. This is for information only.

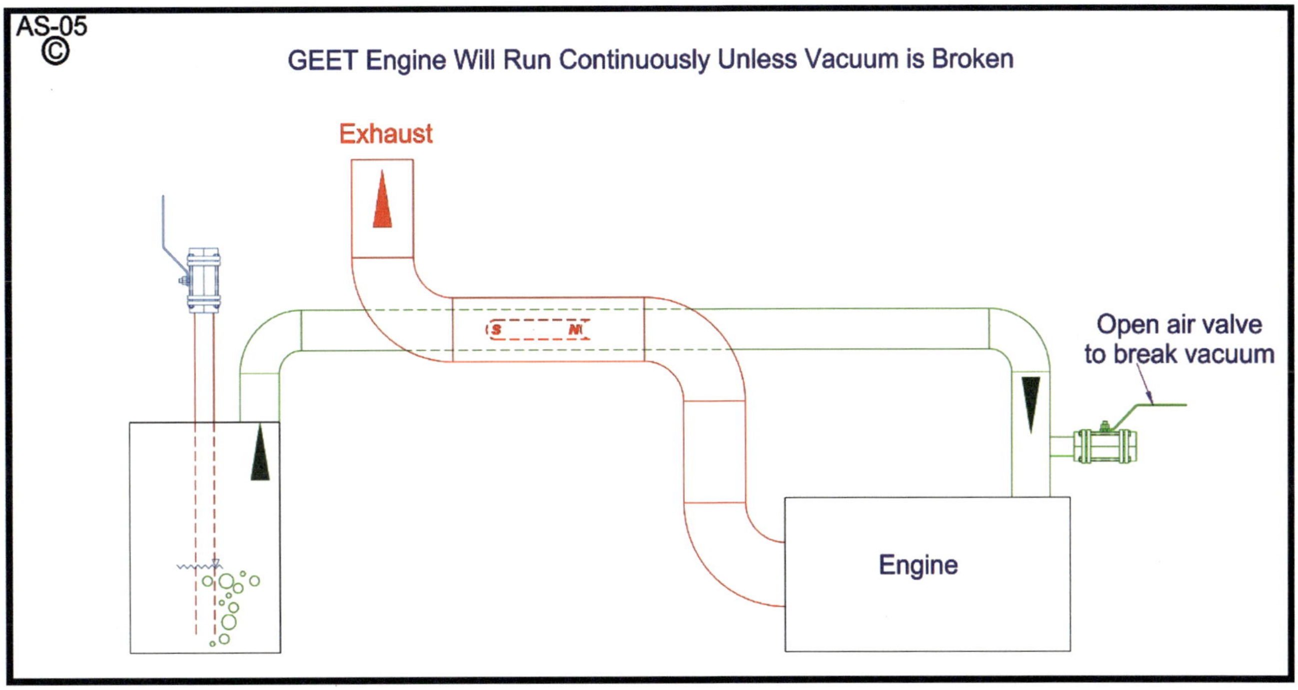

When running a GEET, the ideal vacuum is between seven and eleven inches of mercury. On some older engines, this might be very difficult to achieve, especially on one-cylinder engines. Before we modify any engine, we like to run it first and do a compression check. Sometimes we have noticed fuel coming back out of the intake of the carburetors with factory equipment, and we realized if the valves are not closing completely this could occur.

One of our mechanics in the shop decided to make sure the valves were closing all the way on a small 10 hp engine, so he ground down the valves until they did close all the way. When he put the engine back together and attempted to start it, we could all laugh because he was not aware of the amount of space required so that you could pull start an engine. If the valves close all the way, a 200-pound man cannot pull start the engine.

Have you ever wondered why the old engines used to start almost every single time, and the new garbage we buy today rarely lasts more than one season. They are trying to make engines cheaper and with less quality, no improvement on pollution, and a run time guaranteed to empty your wallet every year.

Production And Personnel Problems

In 1999, we had calls from all over and purchases coming in daily for generators. Many people feared what would happen at midnight on New Year's Day 2000. Companies that manufacture generators could not keep up with demand with regular help, so there were many untrained and unskilled people working the assembly line. Luckily, our test procedure had us fill the generators with oil and do three fuel consumption tests of eight ounces using factory equipment. We continued to do this test until we achieved three test results within 10 percent of each other.

There were so many generators built in such a short period of time using unqualified people so very few were factory tested. One or two out of every 10 generators were returned because they were defective. We bought many of our generators from Sam's Club; their return policy was good, as long as you returned the engine within 30 days of purchase. Most people bought generators and did not start them up within 30 days. Once they did try to start them up, the store who sold it was no longer responsible for replacement and you had to ship them back to the company with your own money.

It soon became obvious to most people who only paid $500 or $600 for a generator they were not going to pay the same amount to send it back to the factory. Trying to find good quality materials is getting more and more difficult. We have become a throwaway society. When I was in my early 20s, we took the time to rebuild a starter and put it back in a car, since it might only need brushes. It would only cost a dollar or two to fix. However, if you go to a dealer now, you get to have a brand-new starter for hundreds of dollars. Come on, everybody's going to make a profit.

Automobile Manufactures Are Scared

Back in 1998, it would have been nice to have a camera on our booth while we were demonstrating GEET at a conference. One such conference was for the Tesla Society and the other half of the building had a conference going on for Ford dealers. You should have seen the looks on the faces of some of the Ford dealers coming through the side door, where we had parked an old Ford Pinto with a sign in the window saying, "This car only get 88 miles to the gallon." It caught all of their attention. About 25 to 30 of their dealers saw the sign and kept coming back to our booth the first day.

One group of four dealers asked how we could possibly be getting three times what their regular cars got, and we explained. Then they avoided us for the remainder of their conference. We could see them coming and going. At lunch breaks, they went out to eat rather than eat in the same dining area as us.

At one point, one of the dealers grabbed a stack of magazines and some flyers we had printed up for our booth guests and ran back to their meeting room and passed them out. My wife saw them doing this from a distance and ran after them. When she went into the meeting room, she announced she would like to get paid for just the face value of the books which were taken from our booth that were marked $9.95, since about 100 magazines were stolen. She even offered to phone the police.

The embarrassment caused all of these dealers to lower their heads and look at their tables. A couple of them jumped up and handed my wife the thousand dollars she requested for the magazines. She took the money, and as she was walking out the door, she said if any of the dealers would like to go for a ride in the Pinto, they would be more than welcome to come back to her booth. From that point on, none of the dealers wearing Ford jackets came near our end of the building. For the remainder of their conference, they walked out the side door and walked around the building to get to and from their cars.

In 2000, I bought a Freightliner and posted on my Internet site that we would have a semi running soon with minimum emissions. We received requests for over a half million vehicles to be retrofitted with GEET. I went ahead and spent about $100,000 preparing the first semi for an over the road test. At that time, I was quite happy as we were getting just under triple the mileage. Then the calls came in along with emails saying they were no longer interested in using GEET on any of their fleets. I found the head mechanic for one of the biggest trucking companies in the United States and asked him why the sudden change. He informed me that Caterpillar, Detroit Diesel and Cummins would cancel all warranties for all of their trucks if they outfitted them with GEET.

I phoned around to every insurance company to find one that might insure GEET and grant warranty insurance to the truck owners. I got the same response from all the insurance companies in the U.S. None of them had enough coverage on their own for a policy that large, but two of them said if Lloyd's of London underwrote them, such a policy was possible. It took about two weeks before I heard from Lloyd's of London. We had an appointment for the last week of August.

The underwriter showed up, went for a ride in the semi, and spent a few days with us. After validating the mileage, the underwriter said it would require some European testing, but he saw no reason why

Lloyd's of London would not underwrite such a company. He did say the fee might be a little steep. He expressed there was another project problem in New York which had to be taken care of before he could address warranty coverage of GEET.

He informed me there were a few very dangerous buildings in New York and could bankrupt most of the U.S. insurance companies if these buildings were to fall due to structural failure. He said the U.S. government was working closely with insurance companies to take care of the situation, and it would be handled in a short period of time. He never did identify the World Trade Center as being the buildings, but by coincidence, it is obvious if they had fallen because of structural failure it would have bankrupted many insurance companies in this country and possibly wiped out Lloyd's of London. Let's not go down that rabbit hole today.

Temperature And Oxygen

When GEET is operating efficiently, the temperature of the air coming out of the exhaust pipe should be the same as the ambient air temperature. However, if you wish more power, you can add small amounts of sugar or salt to achieve more cooling from the exhaust.

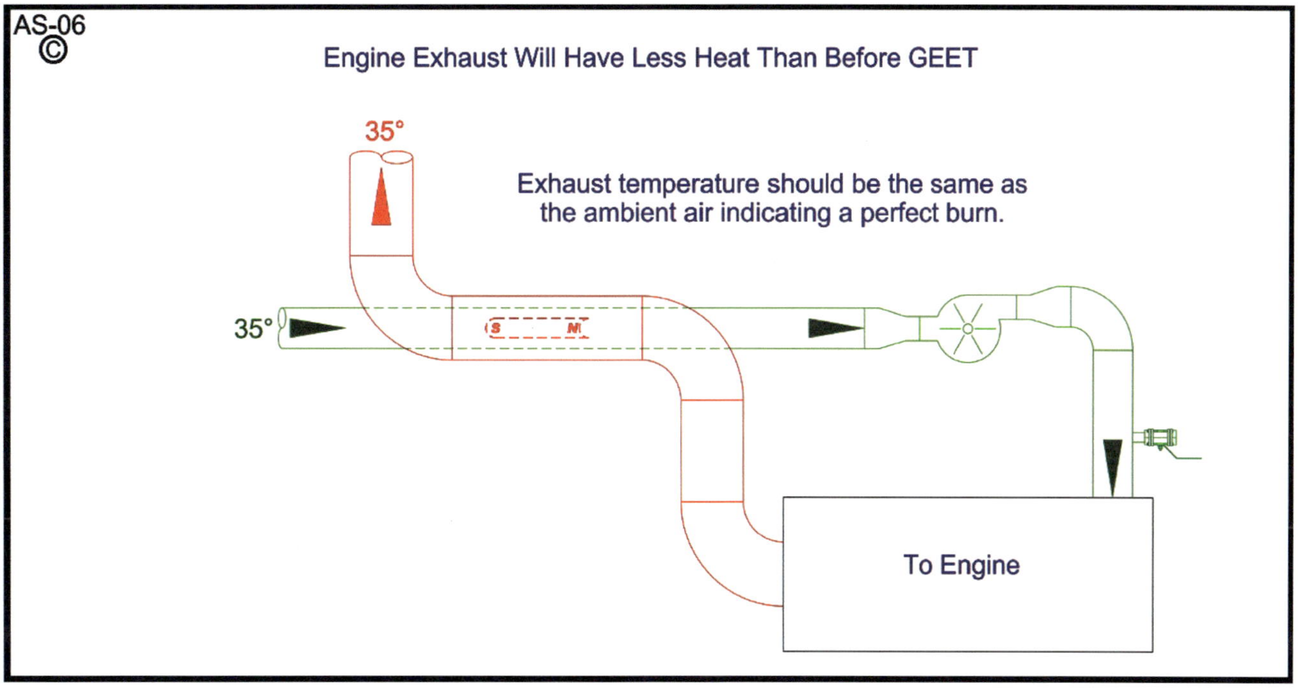

The oxygen level should be the same from the ambient air to the exhaust pipe. When combined, it indicates a good clean plasma reaction. We have had as much as 23 percent O2 from the final exhaust, when using crude oil and water combinations.

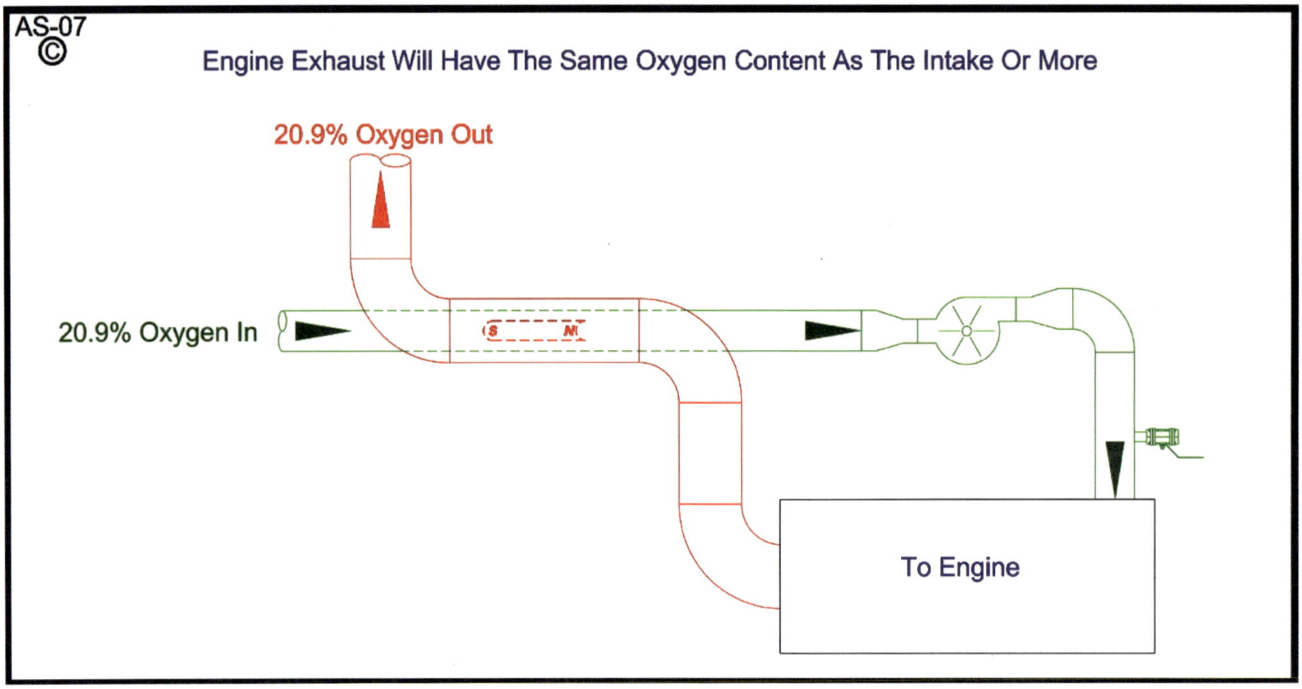

While a V-8 engine can have one or two exhaust pipes, you only need one to operate a GEET. If you want two separate exhaust pipes for different fuels or just for extra power, you may.

How To Stop A Tornado

One of the anomalies we discovered has led us to understand how to stop tornadoes in their tracks. This could be accomplished very simply by having some type of a rocket device either on an airplane or in the back of a police car. It would send a rocket or missile into the center of a tornado and then downward into the ground with a 25 to 50 foot copper tail. Once a tornado is grounded, it immediately releases all the electrical vortex energy into the ground with a large explosion, but it stops the tornado.

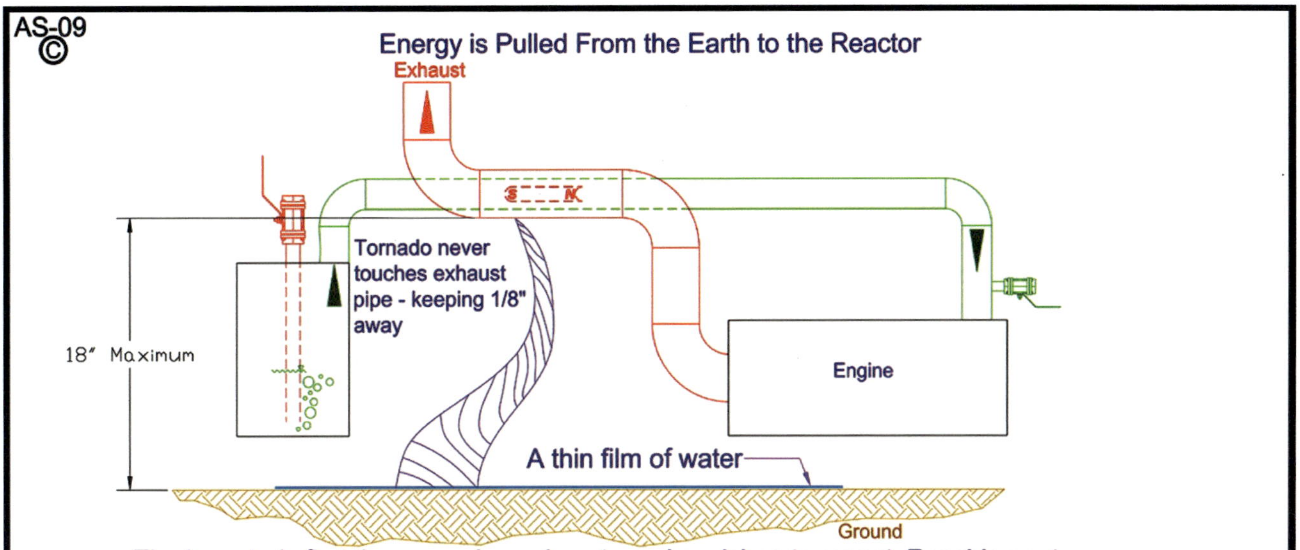

The inventor's first time experience is not one he wishes to repeat. Reaching out while standing on wet concrete, he touched the tornado and was immediately propelled backwards, leaving him with hair standing on end all over his body and with very sore fingers.

A student was running his engine he built after class, and using the standard operating procedures we teach. He put the engine outside the building for its first test run. In case there was a problem, it would not be inside the building. We have never had a student have a problem with an engine, but we do this procedure as a safety measure. On this particular day in Idaho, it was a drizzly day with a lot of moisture in the air, almost raining but not quite. It was just muggy and wet. Since it had been raining for hours before that, our student did not want to go outside and get wet.

The student and teacher went to the back door of the shop and placed the board-mounted engine outside. It had a horizontal GEET application, which was about 14 to 16 inches above the board. They started the engine. About 30 seconds later, as water was running across the top of the mounting board, the student and teacher couldn't help but notice a 14 to 16 inch tall, upside-down tornado. The water was being pulled up to the reaction chamber. The first half-inch rose out of the water, picked up about five inches, and became a funnel cloud. It reached all the way to the center of where the reaction was without touching the exhaust pipe. If it had, it would have grounded it out.

The teacher involved was one of my top instructors and started calling the other shop helpers over to the door to see the tornado machine. One of my employees came running into my office, which was at the far end of the building. He interrupted me while I was typing an email. "Boss! You got to see, this!! It is unbelievable. We have tiny tornadoes at the back door!!!"

I could tell by the excitement in his voice that there was some type of new discovery here, so I asked him what was so unbelievable. He explained everybody in the shop and all my students and secretaries were all out playing with these tornadoes! I do not know how you would react, but I know when I am told there are tornadoes at the back door, I want to go the other way.

I dove over my desk and began running for the front door; as I ran through my office, I realized my secretaries were not at their desks. I glanced around and saw they were also not outside. My shop helper again told me everyone was out back playing with the little tornadoes. I must have had a very puzzled look on my face because, to me, playing and tornadoes do not fit together. I stopped and, very carefully, I went to the back door of the shop to see what the blazes he was talking about. As I approached the rear door to the building, I pushed my way through about 35 onlookers, just to see this tornado machine. I could see the tornadoes about 25 feet away, and I have to admit, they were cute as tornadoes go.

My excitement grew and I went running off to the dressing room. I knew I needed a coat hanger to ground out the tornado, because they did not touch the reactor rod or the exhaust pipe. The tornadoes seemed to know or sense it was a ground and it would kill them. I grabbed a lab coat and yanked it off the hanger. Bending the hanger up, and running back to the engine, I got really close and studied the tornadoes. One would grow bigger and bigger as another one would start up and push it out of the way, and then it would dissipate itself. The energy pulled into the newest tornado and it would get bigger and bigger and bigger. Then one more would start. This continued over and over, but in the middle of each tornado you could see the spinning. It was almost like a mirror on the inside of the tornado with little tiny one and two-inch lightning bolts.

I touched a coat hanger to the exhaust pipe, and slowly moved the end of the coat hanger into the tornado. There was a loud bang, and a couple of seconds later, I was looking at the crowd from about 25 feet away, sitting on my butt, with smoke coming from my fingers and from the opening of my shoes where the electrical discharge did not agree with my body. I also could not help but notice how hot all of my fillings were. My watch was extremely hot, so I took it off. It never worked again. All the other metal on my body was extremely hot and uncomfortable to the touch. Well, I cannot say I discovered lightning, but I can say I discovered how to stop a tornado.

I wanted further proof than just saying I was knocked on my bottom. Everyone came running over to where I was sitting on the concrete, asking if they should call an ambulance. I told them, "Just get my cameras and the tripods, and hurry up. I want to do that again!"

The first reaction from my employees and students was to hold me to the ground. They told me "No." I struggled to get loose. I jumped up and said, "Don't worry, I'm not going to get shocked this time. I need double rubber pads and wooden pads underneath me. I need rubber welding gloves and a welding hood." Once the cameras were set up, I went about taking care of the tornadoes.

About 43 tornadoes later, I stood up and removed the welder's hood. With a big smile, I looked at what was left of the coat hanger, and there were many burns. I went over and turned off the camera, pulled out the cassette, and went and locked it up in my safe. I was not quite sure what to do with this newfound knowledge. Should I offer it to universities so they no longer have to chase tornadoes, putting their lives in danger? Should I contact the insurance companies and tell them how much money I can save them? Should I just sit on the knowledge? What would you do?

Interesting Discovery About Scientific Thought

Many discoveries are made 50 years ahead of their time. Procedures are the key to get the results one is looking for. However, not all findings are truly accurate. Would it be a cold truth or half-truth if I wrote a paper claiming that plants love classical music? Scientists who are paid to do this research have written such papers. If the scientist likes classical music, he will claim the classical music makes the plants healthier. Personally, I believe it is a bunch of bunk.

In 1984, I did a test with my girlfriend who loved rock music. Personally, I like softer more mellow music, like The Carpenters and many others, because I want to hear the words. To validate my test, I bought about 30 plants — two of each kind, one for my office, and one for my girlfriend's. For the next month, her 15 plants seemed to be thriving and so did mine. So 15 plants like rock music and 15 plants like mellow music?

Then the switcheroo. For the next few days, I played rock music. That caused me some discomfort and irritated me at other times, and my girlfriend listened to mellow music which did the same thing for her. Amazingly, all 30 plants showed results of our discomfort and annoyance for the music we were listening to. When we returned to our regular music, our plants came back to life happy. So just because someone says classical music is good for your plant, you must first weigh out how they performed their test and by whom. You see, our results seem to indicate plants reacted more to the feelings of humans than to the style of music played. Interesting…

Let Me See Your Papers

For the longest time, I could not see the real value of the competition held by schools of debates. Then, one young lady, who helped me take care of my son when I was working, invited me to one of her debates. She explained that most of the debates are at night, so attorneys could come in and hear how the students would convincingly argue for or against any issue. She was a good little debater, but as I saw, lacked morals. I only say this because anyone who can stand before me, hold a straight face on Monday and say abortion or death is wrong, and 10 minutes later argue the points of what makes it right, is suspect. Attorneys are nothing more than good debaters. I am not picking on just attorneys. I could be picking on doctors, policemen, or podiatrists.

Next time you go to your doctor, look around for his credentials. You might notice they do not indicate whether he was top of the class or bottom of the class. Maybe the teacher passed him, just to get him out of the class. Attorneys are about the same, but either one can totally destroy your life without really caring since they get paid for what they do whether it is right or wrong. It does not matter. Sounds like prostitution to me.

I did have fun with some attorneys in Utah when I asked them if they had a business license. Not one of them had a business license to operate a business in the state. They all said they had BAR licenses. I checked with the state professional department and asked if people needed licenses in that state. Of course, the answer was quite simple. Anyone, even a five-year-old selling lemonade, needs a license to operate any business at any location, and they need a permanent business license if they do more than $100 worth of business in any quarter. When I asked about attorneys, I was told the state has no legal authority. I wonder who does? Maybe the Queen of England does.

Power And Testing Needed

The lunar cycle does affect GEET and can give you as much as two minutes and 30 seconds extra run time on a five KW generator, if run on a full moon. We have yet to complete the other testing in the area of elevation to determine how many electrons or unidentified fields of energy are at different latitudes or elevations or at prime intersections of lines of force on the earth's surface. However, we have opened the door to phenomenon, which was repeated for 45 days.

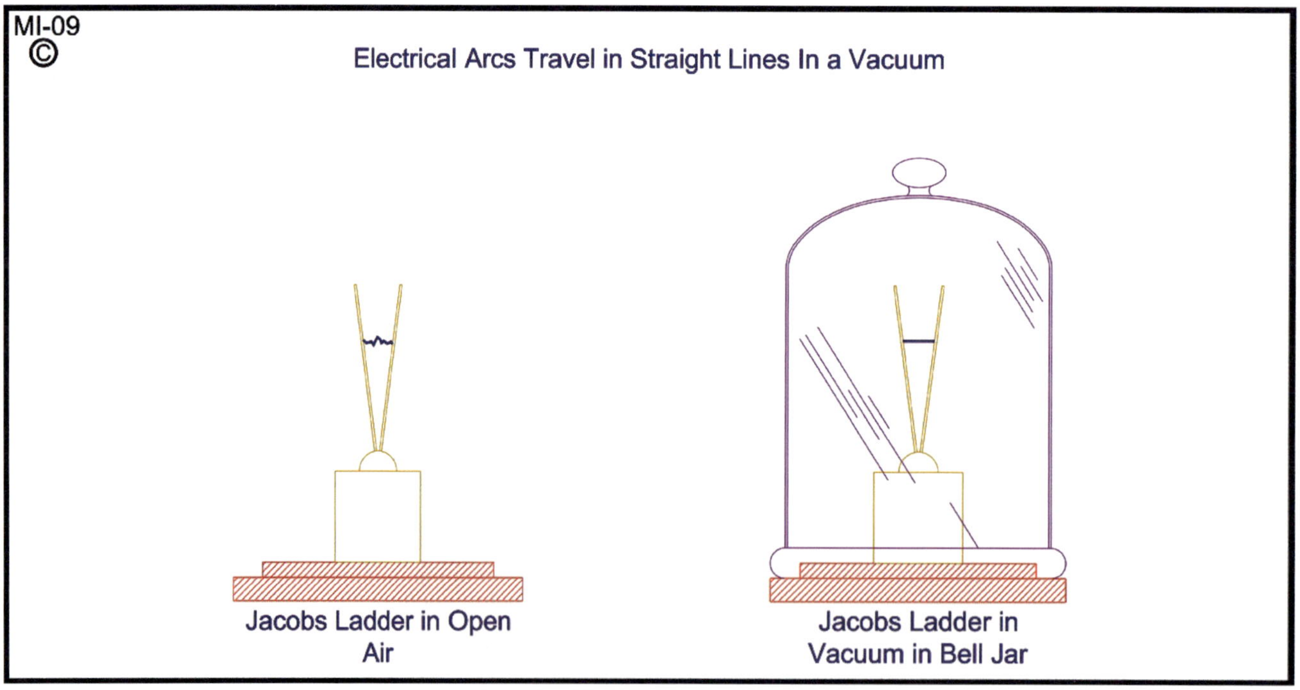

When we were using generators for testing, we tried some horizontal and some vertical, but we had yet to test them in a vacuum. Based on all of the tests, and knowing electricity moves in a straight line in a vacuum, I believe we will find magnified amounts of energy coming from the vertical generator placed in a vacuum. Once this book is complete, we will be able to move on to all of these other fields of testing, and some of you who are reading this book may choose to work with us on any one or two specific discoveries. I will be allowing people to work with me, but I do have a problem working with universities and government agencies. Too often, these entities wish to claim they invented or discovered something when they are not truly the inventor or the party who made the discovery.

Steal From Me, Will You?

I am not going to embarrass all the different universities and professors who tried to steal from me and pretend they invented GEET, because I can prove I invented it. Many people have asked me why I do not accept grant money. The answer is quite simple. Most grants I have seen have said if there are any good findings, then the person or group who gave the grant jointly owns them with the inventor. I do not have partners, and I do not want partners. I do not wish to see any corporation controlling me in this method. I am trying to get GEET out to the world. It seems all companies have a price. I am not a company.

Back in 1985, I had gone to a patent attorney to find out what it would cost to patent GEET. After showing him the engine running on crude oil, he said he would need a few days to put something together. A few days later, he phoned and told me we had a meeting scheduled for 2 o'clock that afternoon, and he had found someone who was willing to pay for the patents worldwide. However, they wanted to see the engine running firsthand. I arrived at the attorney's office a half hour early and set the engine up behind the building. Promptly at 2 p.m., the attorney and his two friends came out for the demonstration. After about 20 minutes, they said they had seen enough.

As they turned to walk back into the building, I told them I would join them as soon as I loaded the engine up. The attorney said not to bother and he would call me in a few days with the results. A few days later, the attorney called me. He said for me to come in as soon as possible. When I arrived at his office, he was still out to lunch but his secretary congratulated me on my sale. Before I could respond, she went on to say not many people could make $10,000 for an old lawn mower engine. I waited for about 30 minutes until he finally arrived and quickly escorted me into his office. He handed me a proposal where I was giving the technology to him, under his control, for $10,000. There was nothing in the agreement covering any royalty or ongoing gratuity for the years I spent developing the engine.

I told him I was not interested in selling the idea for $10,000, and I just wanted a patent. He informed me he had friends in Washington D.C. at the Patent Office, and I would never get a patent. He said if I did not accept his offer he would file a patent for my invention and it would be granted. I would get nothing. When he said this, I laughed and walked out of his office. I knew there was no way he could ever build such a device since I had never taken it apart in front of them.

Validation

Another group of attorneys challenged me as to why it worked and how it worked. They wanted me to validate it from a book. There were no books on plasma in 1984 that I knew of, but these attorneys had to have something in black and white, or they would not believe it. I got up, almost ready to walk out of the room, and I looked to the far side of the conference room. I saw a Bible on the bookshelf.

I must have walked by 300 law books on the way over to the Bible.

"Gentleman, I have the best reference book on earth," I said as I took the Bible from the shelf.

I proceeded to open it to the book of Ezekiel (1:4) and began reading where it mentioned that the chariot came from the north, and I stopped. I put the Bible on a desk.

"Do I need more reference than validation from 2,000 years ago?" I asked.

These attorneys looked at each other and smiled. One of them finally spoke and said it was hard to argue that reference book, but he wanted scientific validation for them to proceed. Had I known what I know now, I could have spent a week doing pictures and explaining the entire phenomenon. This has never been put together until now.

I cannot say all of these discoveries have been here forever. Most of them are common sense or natural phenomenon, but science has ignored them in many cases. Much like plasma — it has always been here, but was only recently accepted as the fourth state of matter. Look at gravity. It was always there, but a man was given credit for its discovery.

That is only my opinion. How important is an opinion? Well it used to be only facts could be used in a courtroom. Repeatedly, I have heard clinical psychologists testify based on opinion and not backed up by any physical fact. Judges make rulings on this. Police officers and many other professions all use opinions. Maybe I am backwards, but I like real facts I can sink my teeth into. I do not want to be told something works. I want to see it work. You can base your life on theories. I want cold hard facts.

When Judge Royal Hansen said I was incompetent and delusional, he based it on opinions of people working for the court and paid by the court. They were to tell the court what it wanted to hear: opinions. I had fact on my side, with thousands of people worldwide who built GEET on their own equipment and were making it work. Even a blind idiot could see the truth, but not that judge.

When I tried to explain my opinions to some of the clinical psychologists, a few listened. Those who listened wanted to know more. The most interesting subject for psychologists was the seven types of dreams I had. I would have thought telekinesis and other subjects would gain their attention.

Small Model Of Our Earth

We have explained the unidentified fields of energy as being a replica of the fields around planet Earth from the axis to the outer protective shields. I believe GEET holds some of the secrets within this plasma field to having a healthy, balanced planet. Because of this, I will have to agree with scientists who have come forward and said GEET is 50 years ahead of man's scientific understanding.

During solar flare incidents, we have noticed an increase in the speed and efficiency of GEET in terms of miles per gallon or hours per gallon. Another anomaly during solar flares is the rod north and south poles reverse, or cease to exist altogether. It was during solar flares when I received phone calls on the same day and they were saying they had no magnetic field on the rod.

All three of these people had tested their rods the day before the solar flare, and all had north up and south down, even when they reversed the rod. I was confused and unable to answer why the field was not there the next day until I started looking at a news video of an earthquake in the Atlantic just off Brazil. There was a vortex going down in the water in the Southern Hemisphere, flowing in the wrong direction.

For water movements going down the drain, the Northern Hemisphere normally spins clockwise whereas the Southern Hemisphere spins counterclockwise. During a solar flare or when the earth's poles flip, this rotation reverses. In my opinion, this occurs on a large scale every 25,800 years. This period of time matches up with several scientific documented cases. We are currently entering into the Milky Way Belt. Once we have gotten to the halfway point, our poles will flip, but not for long — only 25,800 years or thereabouts.

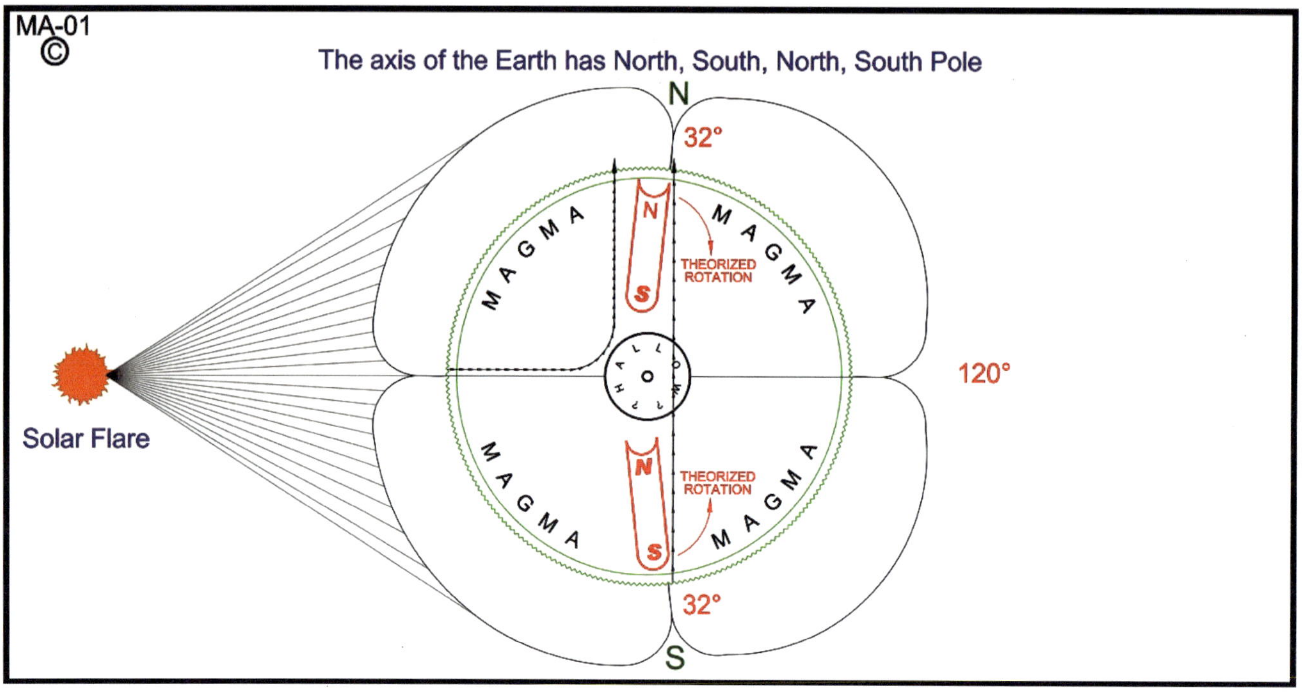

At the last cycle of the Mayan calendar, we began entering the belt of the Milky Way. At this time, we can expect a certain number of asteroids and particles from the belt of the Milky Way to make their way to our beautiful planet. Some burning up the atmosphere and others, like over Russia, exploding in the air. Where and when they land, we do not know, or do we? It would be best for most people to not know, as it would cause chaos.

People would probably panic if they knew an asteroid was going to hit sometime soon. There would be looting, other criminal activity, and many innocent lives taken out of fear, anger, or just self-protection. Can you imagine if the city of Los Angeles or New York received a warning an asteroid would hit in 12 to 24 hours? It would be total pandemonium. There would be no way to evacuate the city to safety. The highways would become bottlenecks and many people would become very violent.

It Is All About Control

I hold back my religious belief, and stick with fact. I do not believe any person should knowingly and willingly steal the property or life of another person without just cause. I do not mean just because I want to. It seems as if other people control the money, control war, control drugs, and our entire legal system. It is not just the government in our country of America; it is all over the world. Governments feel they can just reach into your bank account and take your life savings and whatever else they want as long as you are dumb enough to put up with it. If there is no solidarity and none of you are willing to work together as a team, then you will remain as the workers for the elite.

For years, the governments have controlled what you can and cannot teach in a classroom. It is said the truth can set you free. Obviously, nobody who knows anything about our current criminal system wrote this.

Tasmania Triangle

When the earth is spinning and creating the eight electromagnetic gravitational fields, the aura or auric field around the planet will protect it. A healthy person or plant is most likely not going to be attacked by a virus. If someone has a weak immune system, however, he or she will be attacked viciously. I believe part of this could be detected by Kirlian photography or other non-medical procedures. Of course, we do not want to upset the medical profession as it works hand-in-hand with several government agencies as well as drug companies. You will notice I did not say pharmaceutical companies. I was very specific by calling them drug companies.

Energy spinning into the earth comes from the east. As it hits the surface of the planet and travels through the crust, it reverses direction until it hits the mantle of magma. Then it reverses its direction of spin. Any excess energy which is not required is released in a northern direction of the axis. This display of used energy that becomes more intense during a solar storm is often displayed visibly as the aurora lights.

As the poles begin their flipping, the aurora lights are visible closer and closer to the equator in direct relationship to the current location of the north magnetic pole. Once this flip is completed, the aurora lights will be visible in Australia, Africa, and Antarctica. It will also be during this flip we may lose what is known as the Devil's Triangle, as it will move between Tasmania and Australia.

During a presentation to a group of scientists in Denver in 1997, I mentioned the Tasmanian Triangle. About two minutes later, I saw a hand coming up in the audience, and one of the scientists asked where the Tasmanian Triangle was located. He had never heard of it. Without hesitation, I mentioned it would not be discovered until 2015. Oops!

Many of the things I speak of have come to me in dreams and, at times, I have kept them to myself. About five years after my mom died, my brother gave me a notebook she had written. There were many things discussed and dated on each page. Several events she mentioned had not happened yet. This led to a great deal of frustration because later when these discoveries occurred, credit was not given where the credit was due. This happens to many other inventors. It took many years after Tesla died before he received credit for having discovered and invented many things. The thugs in charge decided to give their friends who had helped them make money, the credit for inventions.

Your Digital Is Vulnerable

One or more of the fields around the reactor have caused significant entertainment. One day back in 1994, Dr. Wood came by for a visit. As he was watching the engine run, he heard a car going down the street. Without turning his head, he identified the type of engine it was and the vehicle it powered. When he turned to look outside, his wallet got within one foot of the reactor while it was running. After the demonstration was over, Dr. Wood was so impressed that he was going to run to the bank and use his ATM to withdraw a few hundred dollars as a gift, since he felt I could use the help. About thirty minutes later, he returned to tell me none of his ATM cards would work. I realized at that point I should have warned him getting a credit card too close to the GEET while it is running would destroy the magnetic swipe strip. That evening I gave him 200 dollars to get home.

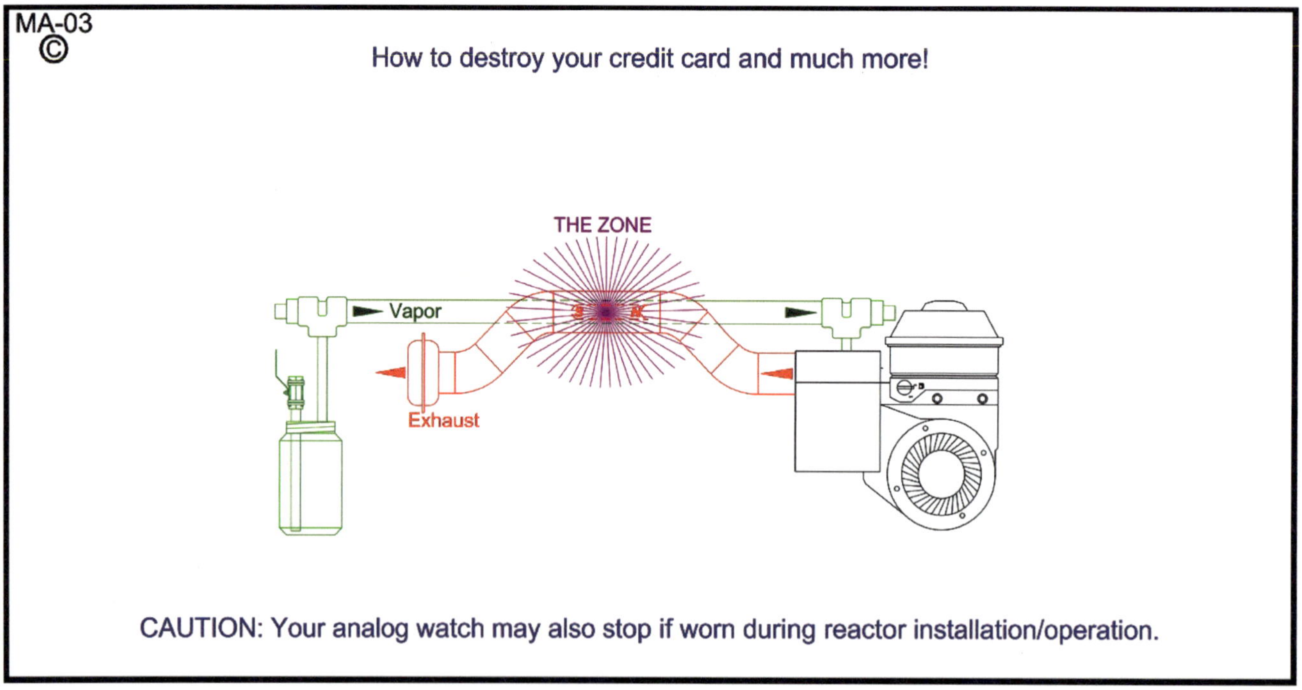

At a conference I attended in 1998, I was moving the rod in my hand within one foot of a digital camera. The red light immediately came on and it went off within a second or two. The writer, Jean Manning, said I had messed up another camera of hers, and the first camera could not be repaired. I apologized for getting too close to her camera with the north end of the rod. She said it was not a problem because she was going to keep that camera in a place of honor on her mantle to prove to people GEET has very unusual characteristics.

One of our students went home and built a GEET reactor for his generator. Once it was running, he decided to preserve the moment on his digital camera. The still picture came out as a bunch of dots. He then used his camcorder, but again he got too close to the engine while it was running. If you look closely at the picture, it appears the copper and steel pipes are moving like pieces of rubber. This phenomenon only occurs in digital equipment if you get too close to the GEET unit while it is running. Please stay at least six feet away when filming, and leave your wallet in a secure place too.

Make A Real Magnet

Speaking of anomaly moments, a mechanic in Utah placed a wrench between the center of the reactor and the intake of the engine by accident. He then started the engine and ran it for about five minutes before he noticed the wrench sitting on the engine. After shutting the engine off, he tried to remove the wrench. It appeared at first as if it was stuck to the engine, and he thought somehow paint on the engine and manifold must have gotten hot and bonded to the wrench.

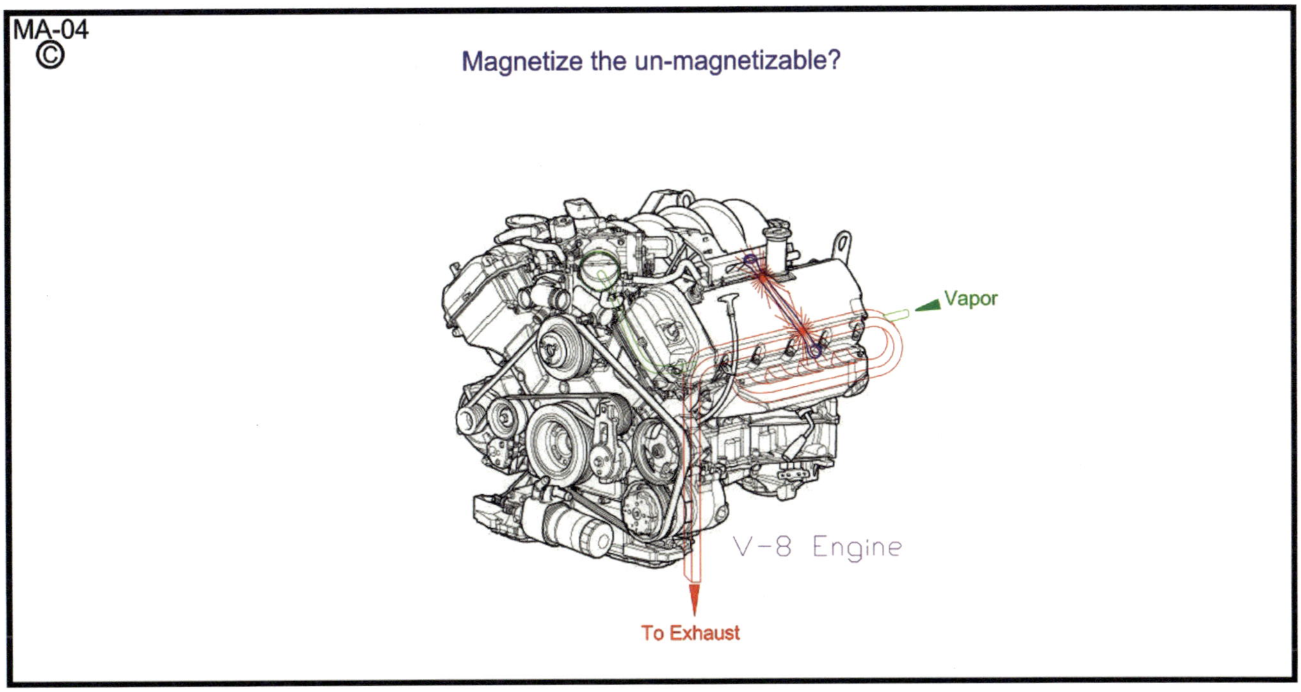

This man finally was able to yank the wrench off the engine and examined it. He saw no evidence of any paint. He dropped the wrench on the rubber platform in front of his toolbox. The wrench immediately jumped up to the face of two of the metal tool cabinet drawers and stuck. He realized he had just turned the wrench into one of the strongest magnets he had ever seen. This is another test we need to do under very close laboratory conditions so we can validate just how strong a magnetic field can be created in a piece of metal or other substance — even ones which are not supposed to be magnetic.

The tool company, from whom he purchased the wrench, brought him a new wrench the next day. They asked him "How on earth were you able to magnetize that type of metal, since it was impossible based on everything we knew?" Since his mechanic had signed a nondisclosure with me, he said he could not discuss with them how it was done. There are certainly many applications for magnets if they can be produced cheaply. Of course, this would only lead to another business being generated by GEET. Moving along, we will now consider all of the other inventions, discoveries, and theories around GEET.

The Other End Of A Black Hole

In December 2000, a friend called me and said there was a website I needed to read immediately. He had forwarded the link by email about a professor from A&M University entitled, "The White Hole Theory." I immediately went to my computer and clicked on the link. I got to about the seventh or eighth page and my wife walked in. I told her she needed to read the article I had forwarded to her email address. She went and read it and understood why I was so excited.

This professor writing his theory had described the operation of GEET quite perfectly. To sum it up, matter is pulled into a plasma vortex if spun at such a speed that it absorbs energy from around it. The energy then comes out at the other end as a ball of white light. Plasma. It took a while, but I did talk to this professor and he was very disheartened by the negative attention the white hole theory had brought upon his life.

His comments reminded me of a paragraph about a scientist in Moscow who was laughed off stage. In a conference when the world scientists accepted plasma as a state of matter, one of the hundreds of scientists who were present got up and presented his theory of how to generate plasma. When he began explaining the plasma could be made in a vacuum instead of under pressure, his colleagues laughed at him. He was not given credit for his superior knowledge and he left the conference.

It is really funny at the time of the conference back in 1983, this little inventor, Paul Pantone, a high school dropout, had built just such a device, without knowing any better. I would not discover until 1994 what I had invented was in fact a self-generating plasma generator. Had it not been for the generosity of Dr. Grant Wood, who gave me a copy of the book entitled "Secrets of a High-Frequency Plasma Generator," published by Boca Raton Press, I might never have known I had invented a plasma generator.

Magnets For Everything

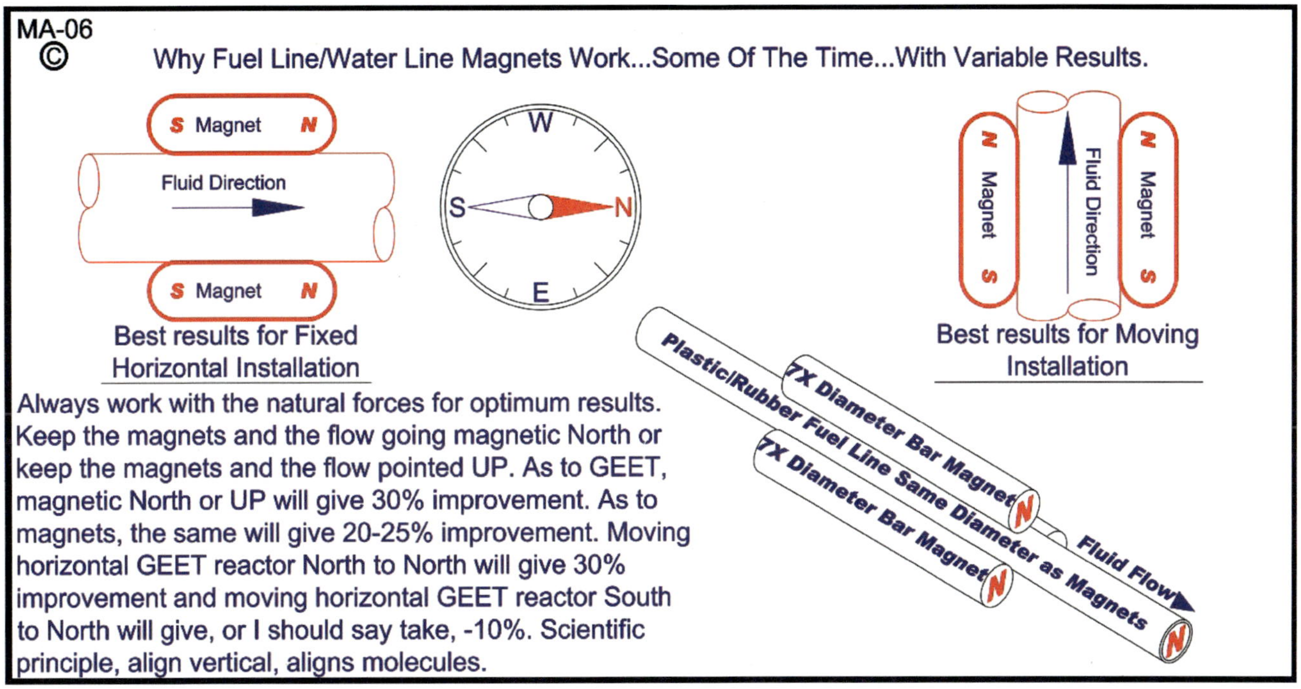

Whenever anyone would introduce me to a new item I did not understand, I wanted to find out if the device truly worked or if it was a placebo. Many people tried to convince me I should try magnets for healing myself, getting better water, or maybe even better fuel mileage. I decided to give it a try. With an open mind, I tried many magnet products. Following their instructions to a T, these products did not work as advertised. However, so many people do swear by them. Why?

I spent thousands of dollars in research and development of magnets to find the truth, or at least part of it. I did find if I put the magnet on a non-metallic fuel line, when it is pointed north, it could result in anything from a 30 percent increase to doubling the mileage. If it were on a horizontal plane and a non-metallic fuel line, mileage would go down when the fuel line was pointing south and only go up when the fuel line pointed north. We found no noticeable difference using magnets on metal fuel lines.

After a few years, I did realize that magnets do work, if only under exacting conditions. None of the other companies selling magnets knew this. The correct orientation to north and south seems to be the most important condition.

I found an owner of one of the largest multilevel magnet companies in the country. I told him I had figured out the secret to get 100 percent satisfaction from his customers. The guy said he really didn't give a damn. He sells his upper level magnets for five dollars each, and by the time they get to the public, some people were paying as high as 250 dollars for a pair of magnets. They had about a 20 percent chance of working as advertised, and again he said he just did not care. He made his money on the front end. There are so many companies marketing their products even if they do not work. Basically, they were selling the sizzle, not the steak. Yet, they are allowed to exist without any resistance from the government or oil companies. It is pretty obvious if something has a chance to hurt the oil companies profits, it will end up with a black eye and no one will buy the product.

Make GEET Go Away Or Steal It

Look at all the mileage we can get from an electric car. If you really want to get sad, go to any website search engine and look for "cars that cannot be sold in America." There are many of them. Most get very good mileage, but they would hurt the oil industry and/or the utility company profits. You cannot be politically incorrect in America or they will shut your doors.

If you are a writer, journalist, or TV reporter, the chances are someone controls you. You do not dare say the wrong thing or you may lose all your credibility. If not, the IRS can seize everything you have so you do not have the funds to fight them. Often I have been on radio shows or news programs and was later told they can never have me on again. There have been five movies shot about me in the United States, yet they cannot be shown on television. You have to ask yourself 'Why?'.

In 2014, History Channel South America sent a film crew to my home in New Mexico. I spent the day being filmed and running engines as I would for any other TV special. I really did not think it would ever make it to television. I had requested I do the final editing before the release of the film, and they agreed. In November 2014, the History Channel put the show on the air in Spanish and Portuguese throughout South America. I was flooded with calls, emails, and letters. Most thanked me for what I was trying to do for all of humanity. I received more response from the 35 to 40 times the show played in a four-month period than I have with all of the Internet work I have done in the last 15 years. People from Argentina and Brazil and the other South American countries have shown so much support and desire for a clean planet as well as putting their people to work. I wanted this reception in my own country of the United States. For a while, I had felt this was more like the Divided States of America. I now realize this is not the case and we should call ourselves the United *State* of America.

When the king of Saudi Arabia discussed GEET with one of his sons, who was the prince of Kuwait, back in 1999, he set an appointment to meet with me in my home in July. When the prince arrived, he made it abundantly clear as we greeted he did not intend to shake my hand, as I was an infidel. I demonstrated the engine for him and his entourage, and then turned down his offer to buy GEET. I then promised him as long as I was alive and my family and dealers were all safe and unharmed, I would not release the plans on how to run a car or a generator on water.

The prince understood I was helping his people and all of the people of this planet by holding back my technology until I could teach the Arab nations and many other nations how to grow healthy food, desalinate water and replace the income they would lose. I do not wish to bring harm on any group or nation, but to educate them so they may all have clean air, fresh water and energy efficiency far beyond what we have today. This task can only be done carefully and systematically to assure each country and people a better lifestyle for their children and their children's children.

Turning down the buyout offer from this Saudi Arabian prince and OPEC was the beginning of a new hell for me, and a bumpy road for many others. Once the prince left my home, he had his limo drive over to the office and shop of GEET International LLC, or GEET LLC. In the shop, there was a man pretending to be an attorney, Robert Duden Beer, acting as legal counsel in-house for GEET. At the time, he failed to tell us his license was suspended. Mike, the president of GEET LLC, hired this attorney to help expand GEET into the rest of the world. There were also three shop helpers Mike hired to assist in the building of generators and automotive products.

As a sidebar, before going on with the invention, I must clarify this information. I was not a member or shareholder of GEET LLC. I operated the GEET research and development work, under contract with GEET LLC, which was a management company. Now before GEET LLC, there was a company called GEET Management LLC. This company closed when it was determined the managers/owners had been stealing money from the company for personal gain.

GEET Management did not have the legal right to sell ownership of the GEET patent, nor any right to take anything beyond their paychecks from the company accounts. When local police were called regarding the attempted theft, they refused to take a report. The two men running GEET Management quickly fled the company and left it broke. A new contract had GEET LLC responsible for marketing. Mike was appointed the president of this new company.

The day after the prince visited their shop, Mike phoned me and asked why I would turn down three billion dollars. Mike, who claimed to be a good church-going Mormon, said such an offer would really help everyone. It was many months before I realized who everyone was. I was frantically trying to get money raised so that we could have an abundance of pre-manufactured equipment ready to go out the door. The orders would come in, generators would be bought and modified, and then go back out the door.

I was reviewing all the UPS books to see who had generators shipped by them. It was not until January when I had all I could take from this management company. There were just too many weird things happening. One businessman from Australia who said he wanted to mass-produce generators in his country was supposed to make a quarter million dollar transfer just a few days before Christmas for those rights, and it didn't happen. There were also many threats by disgruntled people who had not received their generators.

We were expecting checks totaling about 1.5 million dollars from several other countries. None of them came in for deposit. Here it was just a few days before Christmas and no money. On New Year's Eve, Mike came out to my house and brought a bottle of booze so we could celebrate the New Year. He dropped the bomb on my wife and I "The company would go belly-up in a matter of weeks."

My wife packed up all of her personal things and called a friend to move her from Utah to Colorado as quickly as possible so she would not lose any of her personal belongings. She also took with her a copy of all the records I had in case they were confiscated for any reason. This is for another book called, "Terror in Teller County."

I have resisted for many years using a computer, as most of the information I gleaned was not validated or substantiated by any "scientific proof!" Out of desperation to save my company, I phoned a man who builds computers. He wanted a set of my plans, so I asked him if he could come over and give me a computer and a crash course on how to use it, and I would give him a set of plans. Within two weeks, I had a new website set up and hidden so I was prepared to take over the company if need be.

Out of nowhere, my computer man and his attorney made an offer to take over GEET LLC and run the company properly so my name would be kept out of it. I would be completely free to do the research, development, and teaching, as I was under contract to do with GEET LLC. Mr. Stein, the attorney, prepared all of the documents, and he and his partner filed with the state of Utah.

On January 15, 2000, I went to a beauty shop and said, "All the hair goes, and I want to look like a redneck." I then went home and put on an all black suit right down to the tie. I had prepared a seven-page document of all the criminal activities being done by GEET LLC. I then proceeded to drive over to the office of GEET LLC and backed up to the front door.

When I opened the front door, no one recognized me. Here was a dark-haired guy with no white beard and an all black suit wearing dark shades, demanding they all come forward. As they did, I handed each one of them a packet of seven papers. I gave them a choice of getting off the property within 10 minutes with personal effects, or I would a U.S. marshal arrest them. Mike and his wife ran to their desks and started to open the drawers.

"Personal items only — no documents," I said.

The attorney said he had personal documents he had to take along to clients.

"Your only client is supposed to be GEET," I immediately responded. "What other client documents would you have in our office?"

He stuttered, grabbed his briefcase, and ran to the door. The others followed. Two of the five left their laptops right where they sat, and they all drove off quickly.

I had help show up and we moved all documents and computers to my home. I began the arduous task of going through all the emails the company received. It was mind blowing. I found the new brochures for the new company called heat, H.E.E.T, or HEET. The attorney not only was heading up the new company, but he was also handling divorces and other legal problems while charging the money/fees to our account for his time.

The biggest surprise was finding all the people I had previously fired working in the shop. Mike hired back every enemy of the company, to learn from them how to steal more. He thought he would not be caught because his exit strategy was to file bankruptcy. Then he would simultaneously file for back wages and claim ownership to the patent and all patent rights in exchange for all the money he did not have coming.

Within their computers were letters they had not deleted to many of these investors. During the last 30 days he had told all of them GEET would never succeed. He said the new company, which would be formed by the end of January, would be happy to accommodate their needs if they could just hold on to their monies totaling one and a half million dollars for a few extra weeks.

It took about three weeks to get things turned around, and I was heading in the right direction. In February, I was reunited with my wife in Colorado. I thought we were only a few miles away from her "best man and bridesmaid" who pretended to be our best friends. Boy, was I wrong. I had done no more than move from one hellhole to another. Greed and power led them to want to be the new GEET Incorporated or Corporation. Their plan was to get me removed by alleging I was trying to kill my wife.

By June, new documents had been prepared for the corporation. Prior to incorporating, our new attorney went to Wells Fargo and opened a new corporate account with no documentation except a generated letter he had produced. He spent all the money for traveling and entertainment. Once incorporated and before any other contract was signed to operate the company, they were pulling in money from product sales and selling stock. Just before they decided to have me thrown in prison for the attempted murder of my wife, they presented me with a document which was supposed to protect my wife and me. It was a simple trust agreement making Mr. Stein the sole trustee of the R&D and all other patent and copyrights in the event anything happened to both of us.

In June, since I refused to sign the new trust agreement and they had refused to give me any contract for approval, I asked them to step down and return my documents. They immediately began falsifying reports of my harming or threatening to harm my wife. It almost worked. When I knew something was up, I left Colorado and went back to Utah. I did not know my wife had been drugged, and would not know it for at least a month or two. What was done to me was done maliciously and intentionally to take this technology, kill my wife, and put me away for life. As you can see, I am very much alive and single.

Every time I moved, I lost more and more of the documents that supported all of the GEET tests and validations. I was becoming very proficient at recreating many of the documents. However, some can never be recreated. Several of the agreements I had were with inventors who had turned technology over to me and were now dead. I have no proof I purchased their inventions. Since the assignment of all those rights for the different inventions were transferred to me without valid, physical proof, I have nothing.

This is why it is so imperative for me to get this knowledge out to as many people as I can. By getting this knowledge out and in copyrighted form, I am not giving away the knowledge to be used for any commercial application. It is to be shared and not hidden or put up on a shelf like the oil companies and governments would like. This problem is not just limited to them. We all know the General Electric and Siemens of the world would be very upset if they found out we could make an x-ray or an MRI unit that costs less than five hundred dollars.

There are ways to run your car and you would never need another damn drop of gas or diesel again. These are just a few of the things, which can be done. I am trying to be politically correct; simply by educating people, so valuable knowledge does not get lost. Still, there are many things mentioned here giving you the connections to other fields, and which again may not be politically correct.

Medical Eye Openers

When we began doing the tests with Met-Chem Testing Laboratories in Salt Lake City, I had no idea I had created a new type of x-ray or MRI machine. It was just one of the side benefits of building a GEET. Since it does not put out alpha, beta, or gamma rays, this could possibly change the x-ray business and magnetic imaging for the next generation.

ME-01
©

Exploded view of the Hose Clamp

The hose clamp is stainless steel, holding together a split carbon steel exhaust system around the reactor.

Hospital grade X-ray film held near the reactor area exposed the film except at the hose clamp. It has been determined the energy was pushed from the reactor causing the image of the hose clamp to be exposed onto the film.

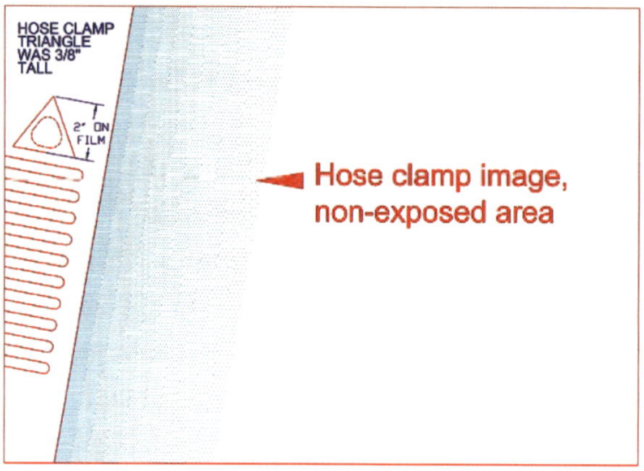

X-ray Film ~12"
away from reactor

ME-02
©

Two different rays detected

The fingers showed only the soft tissue from the near side of the film due to the protected rays

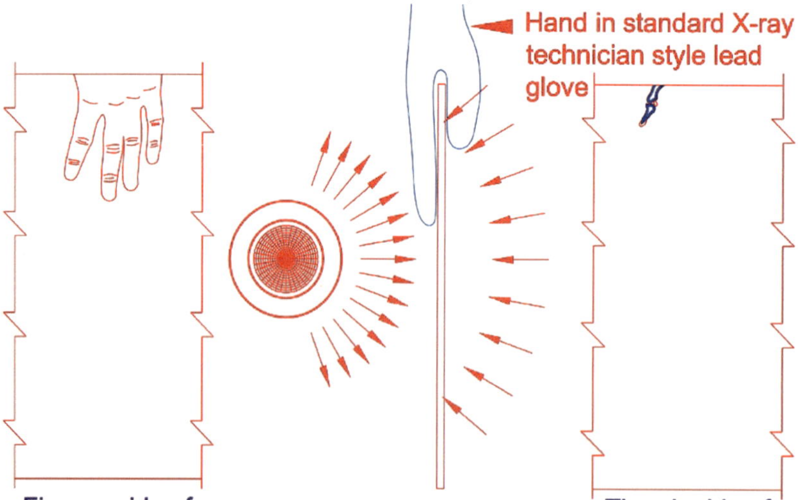

The thumb showed excellent clarity of the skeletal structure from the off side of the film due to the incoming rays

Fingers side of X-ray Film | Reactor | X-ray Film | Thumb side of X-ray Film

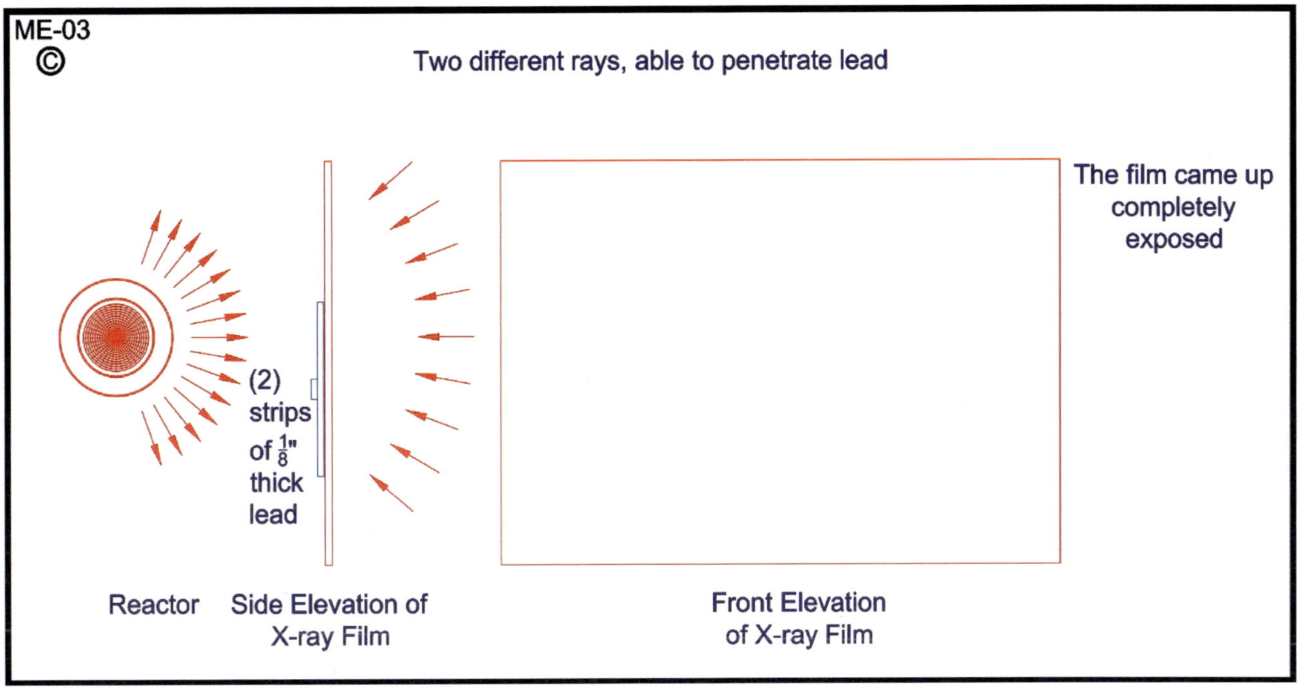
ME-03
©
Two different rays, able to penetrate lead
The film came up completely exposed
(2) strips of 1/8" thick lead
Reactor
Side Elevation of X-ray Film
Front Elevation of X-ray Film

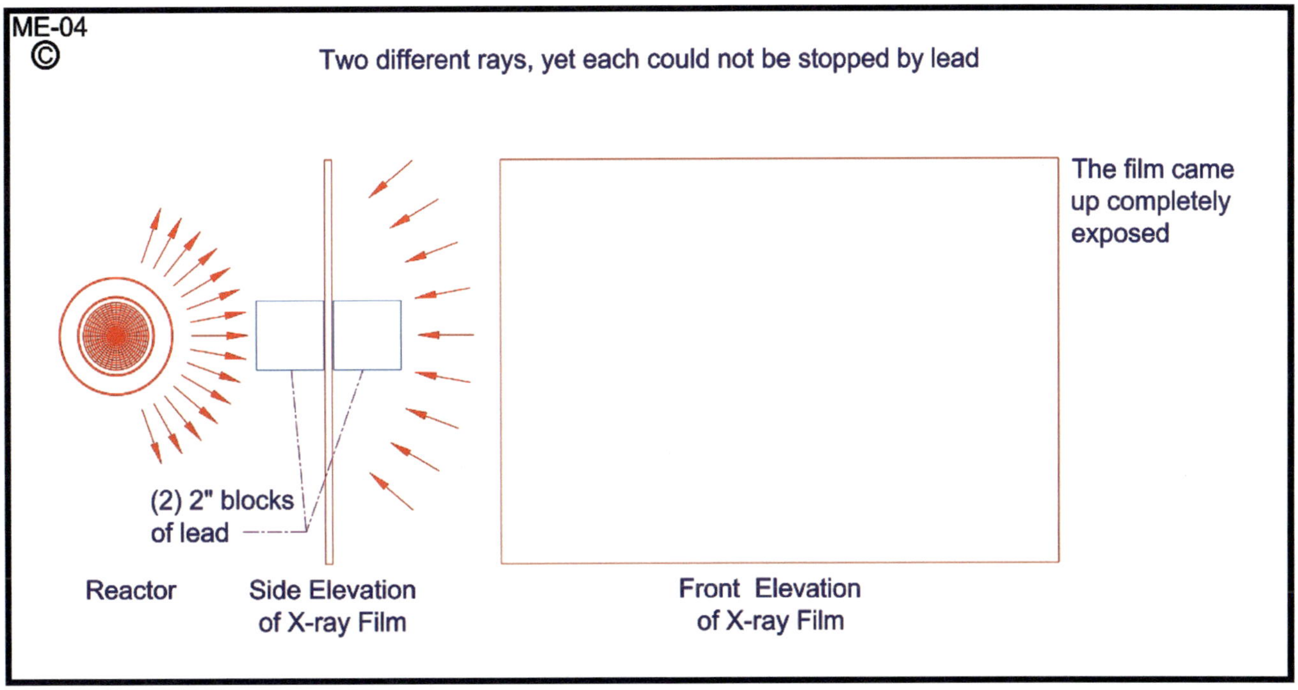
ME-04
©
Two different rays, yet each could not be stopped by lead
The film came up completely exposed
(2) 2" blocks of lead
Reactor
Side Elevation of X-ray Film
Front Elevation of X-ray Film

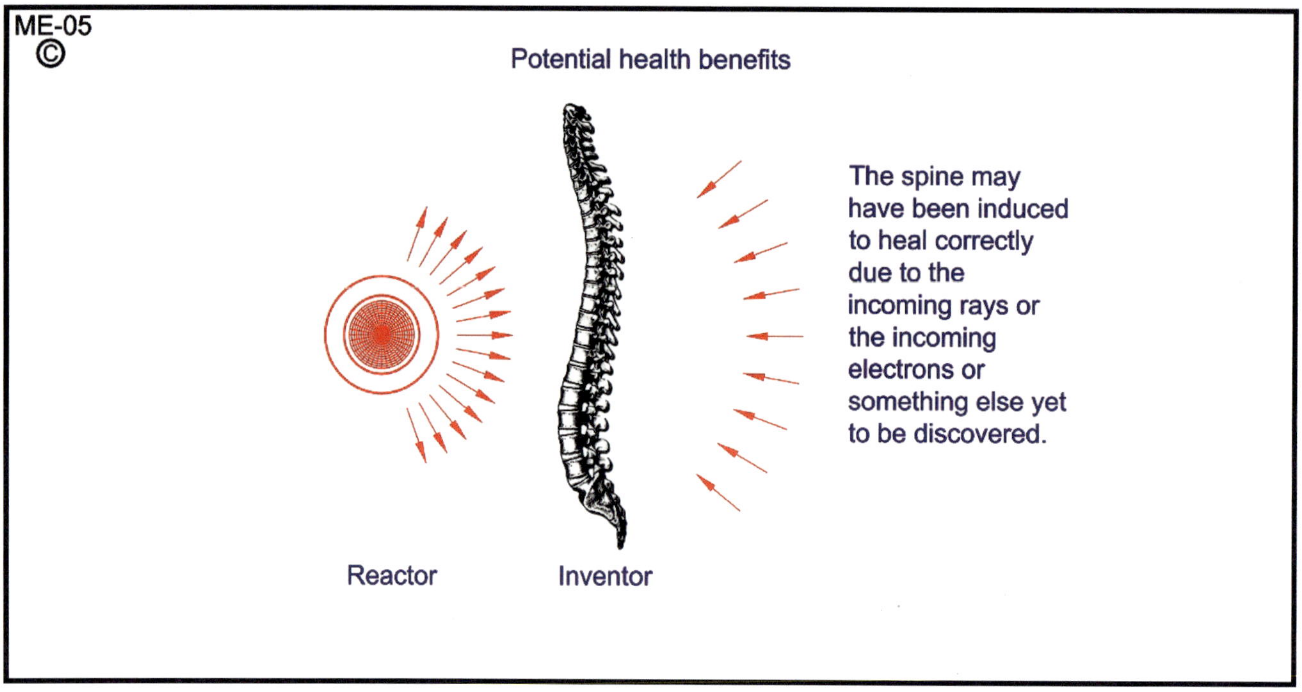

Sometimes when a product comes out, people put up resistance or laugh about it. Many people did not like the VW Beetle, as they felt cramped. It was a small car produced at a time when everybody wanted big, luxury cars. It took a while, but now look around. You sure see many small cars. Again, I have to go back to my first question. If they could reduce the weight of the car by 50 percent and make it so much more efficient with all the electronic controls and fuel injection, then why are we not getting better mileage?

If you ask your congressman or senator if they would consider sending back any contribution from an oil company, I think you would find you just lost your communication with your senator or congressman. It is all about greed and control. Greed happens in all of society, but some is not so noticeable.

One of my dreams is to build a specialized GEET unit large enough for people to get into. If GEET can put a healthy aura around a piece of glass or metal, what kind and quality of an aura could you put on a human being? I would like to know, but just one unit is going to cost about $500,000 to build, and I do not have the money to do it. I have looked for angel investors. Some promised they would back me, but when pressure was put on them, they all backed out. From just how high up do these threats come?

If I do not build this specialized, human-sized GEET unit, then someone else may and I can help. There will be someone who wants good health for themselves or a loved one.

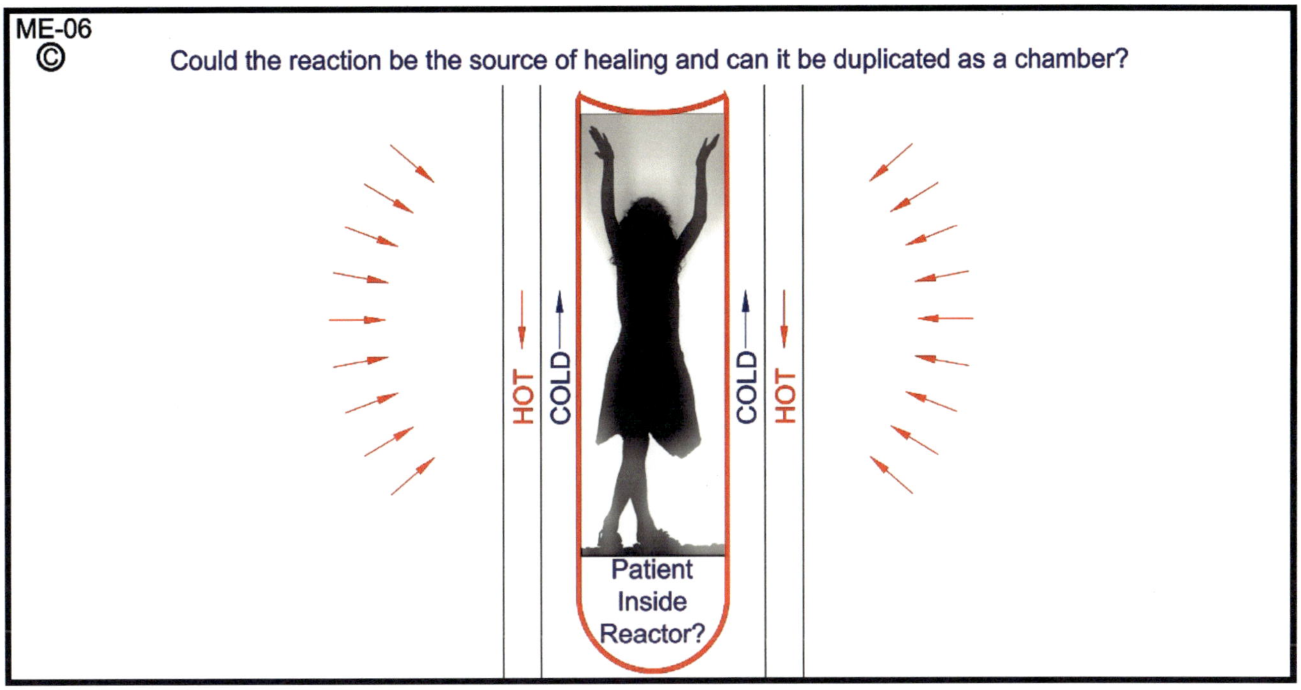

Imagine such a machine where you climb into a glass chamber with hot and cold going in opposite directions. The resulting creation of an electromagnetic gravitational field might cause a healing effect on your body. I for one would like to try it, and I hope the money will be here soon to do it.

You Can't Stop GEET

Back in the 1980s, many people bragged about their new fuel injected cars. They talked about good performance and clean exhaust. Sorry, folks, it was all BS. Oil companies and car companies were fully aware carbureted vehicles could easily utilize a technology like GEET. I am sure the Fuels Research Institute, which was funded with your money but operated by oil companies out of Tulsa, was set up to shut down inventors of fuel efficient devices made by: Fish, Fisher, Pogue, and even Tom Ogle, etc.

The F.R.I. examined many GEETs. The problem they had with GEET was you could not change the formula of gasoline and mess it up as they had with previous high efficient carburetors. This is why they had to close down the Fuels Research Institute. I do feel rather proud of myself for that one. If the best the oil companies could come up with was trying to sabotage the fuel as they did with other inventions, their modified fuels could not do it with GEET. Some people just want a free ride in life, at your expense or mine. My message to them is to get a job and pay your way through life. Every day I read about someone who has been messed up over false bad reviews posted on the Internet. Sometimes it is as simple as a person sitting in a restaurant and telling the waiter to make him happy or he will destroy the company's reputation by posting it on Yelp. I call it blackmail. I do not like the idea of big brother and everybody having a camera on every corner, but sometimes it is the only way to prove you are innocent or the other person is guilty.

This method has been tried with me a few times. "Give me a generator to test that I get to keep, and I will help you promote your company on my website." Some of them tell you if you do not do this, they will destroy your reputation by claiming you are a charlatan or just an outright crook. My response has been "If you want one, then buy one." My other response is "If you want to build your own so you can see nobody put special parts in it, then buy the plans and build one yourself."

It is pretty sad some of these people who claim to be Internet reporters are just plain nuts or extortionists. Can you imagine going to a big restaurant, a nice one where the average meal is $100, and thinking, "Serve me a good meal, and if I like it I'm going to go out and tell people how good you are, but I don't want to pay for the meal." This is the kind of audacity some of these people have.

You Just Need To Find Your Niche

As a young man in my teens, I learned a very valuable lesson from my mother. She drove me down the street one day when I was 12 years old and asked me look at all the houses and tell her what type of work each house needed. When we got to the end of the street, I said I did not see anything other than cluttered yards. She struck me with the back of her right hand. Then she screamed at me, "Are you blind?" She then turned the car around and stopped in front of every house until I found something at every house that needed some attention.

At the first house, before she could hit me again, I said the trim needed to be scraped and painted. For twenty houses after that, I found something wrong with each and every one of them that needed repair or maintenance. When we got to the last house on the street, I began to panic. It was meticulous, not even a blade of grass hanging over the concrete. I covered my head to protect it. She laughed and said, "Okay, where's the address?" I looked around and could not see an address on the house. I could not see the one painted on the curb, because of a car parked in front of it, and I told her this.

"Why don't they paint the numbers in the middle of the concrete in the driveway where people don't park?" she said then. "Wouldn't that be of service? There are many times I have seen ambulances or policemen coming down the street with people standing in front of their homes waving them down instead of being inside helping someone. If the numbers were there, they would be visible to everyone. Help could arrive quicker." This made sense.

Thanks, Mom. I painted a lot of numbers on many of the aprons of driveways in many states. Back in 1967, I was caught painting numbers on driveways by a Beverly Hills cop. I was walking away from a door where I had just told the owner as a courtesy that we were painting numbers in the driveways so if they ever needed an emergency vehicle, their home would be easy to spot. I handed them a small piece of paper which said 'this was a gift from us to show police departments and rescue squads how a community should work together. If they wished to make a donation, it would be graciously accepted.' I also asked them to let the paint dry for at least 15 to 20 minutes before taking any chances of driving over the numbers. Donations were very good.

The Beverly Hills police officer stepped to the front of his car, crossed his arms, and asked me to come over to talk to him. He asked me what I was handing out to these homes. I handed him one of the small fliers.

"Do you have a solicitor's license?" he asked.

When I said no, he explained in Beverly Hills if someone is going door-to-door selling subscriptions or gathering donations, they must have a solicitor's license.

When I explained what I was doing was for him and for all the rescue squads, I asked if he would like one in his driveway. He thought about it for a minute then proceeded to retrieve a wallet from his back pocket. I really thought I was going to get a ticket, but to my surprise, he pulled out his wallet, opened it, and pulled out a twenty-dollar bill. As he handed me the money, he said "I like what you are doing, but to get it done by Monday before I get back on shift or I will have to give you a ticket." As he was getting back in his car, he smiled and said "Have a nice day." In America today, things are much different. So often, I read about children selling lemonade and getting busted for doing so without a license. The little girls are now giving away lemonade and accepting donations. They are now making some good money.

In some communities, the city leaders allow the laws to be modified so you cannot even have ice cream trucks drive down community streets. Some of these laws were made to be broken, just as GEET does with the laws of thermodynamics. You can call them the laws or flaws of thermodynamics, but they surely do not apply to plasma, the fourth state of matter.

Fuel Injectors Are Not The Answer

Let's get back to the "great" fuel injectors. After I had 500 fuel injector blocks built to use for GEET kits, I went out to buy 1,000 fuel injectors in Salt Lake City. Very few shops were willing to even sell me more than a set of eight or ten, so I bought the ones I could. Since they were not identical to the ones we had used in testing, I decided to do a little bit more testing with these new injectors. I was blown away with the results, and you will be too. There was as much as a 50 percent difference in the amount of fuel delivered that one injector put out over another, with no way to adjust, balance, or control the fuel flow.

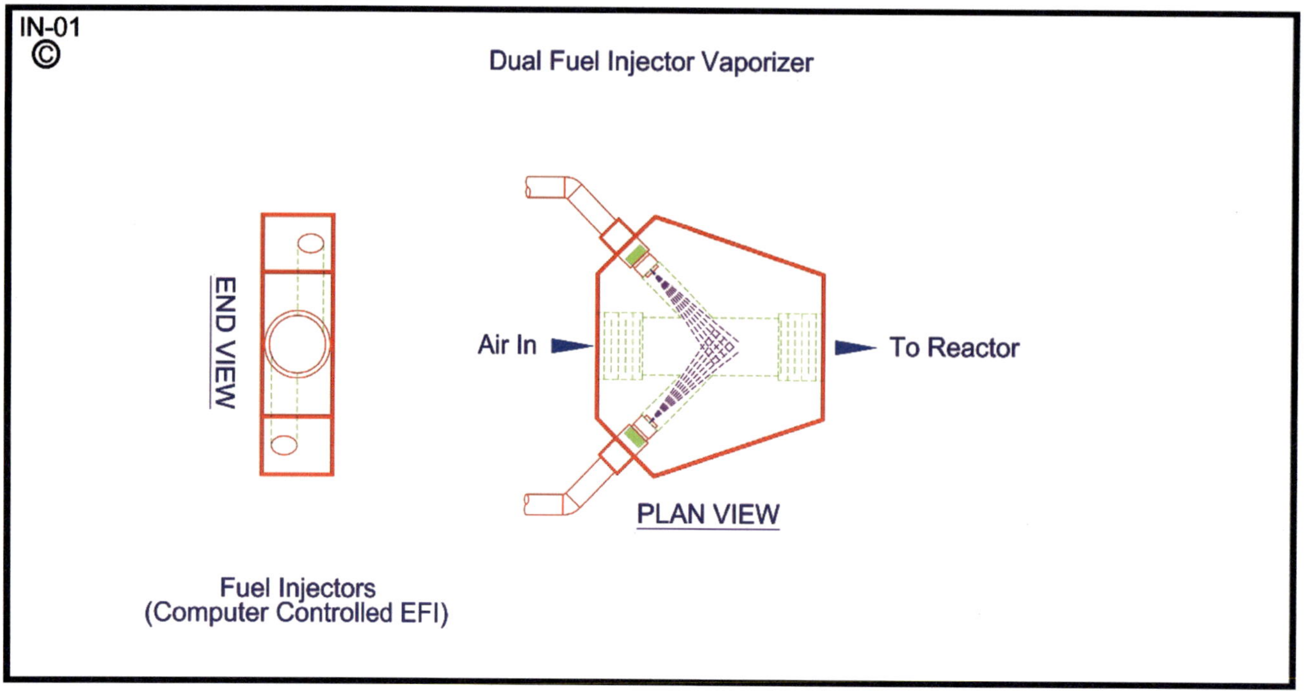

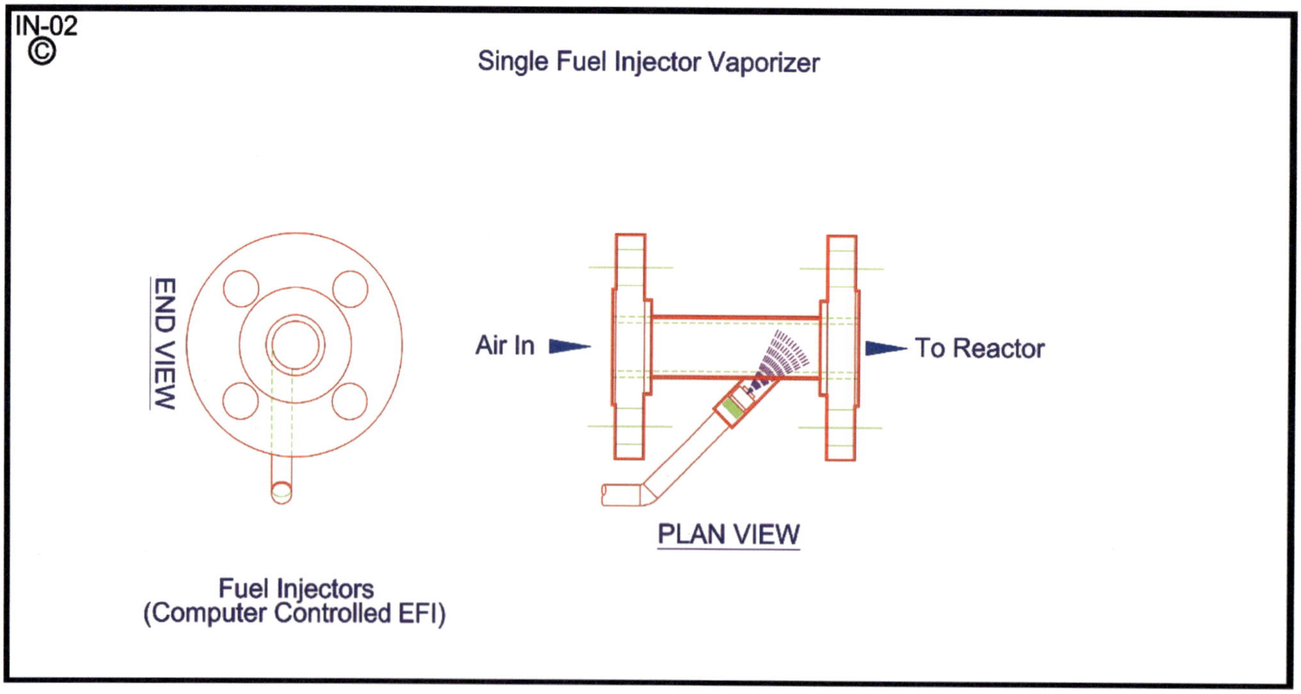

Please, can anyone out there tell me how this could make sense? A mechanic's reason for this discrepancy was we did not know how to test like the EPA or car companies. Running an injector hooked to a wire connected to your car and turning it over for one minute should provide similar results for each of the same set of injectors being tested, not a 50 percent discrepancy. If you want your fuel-injected vehicle to run at peak performance, each injector should deliver the exact same amount of fuel to each cylinder.

We are currently working with a team in an undisclosed location to do such testing. When the results are in, there may be a new company selling balanced or balanceable injectors as a set. Until the tests are done, we will not know if the current computer settings in your car will allow for such efficiency since companies like General Motors and Honda and all the others keep their programming secrets from the public. The computerized sensors in new cars will stop the car from running if your vehicle is operating too clean. Ironic, isn't it?

Back in 1985, after my first TV news broadcast, I received over 100 death threats in one day. When I found the reporter and told him of these death threats, he simply responded by saying it was not newsworthy. I asked him what he meant. He said the television station had notified him that even if I could walk on water, I was never to be on the news again or he would never work as a reporter again.

After buying all the new injectors and discovering these big discrepancies, I decided it might make more sense to go to the wrecking yard and buy a bunch to see if we could come up with some way of making them adjustable. Boy, was I rattled when I got to the wrecking yards. They said they could no longer sell injectors to the public. Why? An injector company was paying them to make sure the used injectors would be destroyed. If any wrecking yard at that time were to sell me, or even give me an injector, they would not be paid the six to eight dollars per injector for destroying them.

This Is How We Got Here

It has taken many years to understand why Ford accepted the gift of his first foundry. Henry Ford was hoping to put steam engines in his cars and knew this would upset the Rockefellers, J.P. Morgan, Carnegie, and many more. They all got together and decided Standard Oil should donate the foundry or the money to buy it for Mr. Ford under the condition that all cars he makes would have gasoline-powered engines.

Knowing what I know now, I can see how this makes sense. If you are making every single part, then no one has control of your production. However, if you are buying parts from anyone who can be bought out, you may not be able to get the parts you want if you are offending someone with more money or power. You have to remember back in the late 1800s gasoline was a waste product burned off at night since it had no practical use. If Ford could sell enough cars with gasoline engines, and oil companies could sell gas at their new gas stations they own and operated, they would make scads of money. Who cared about pollution? Their grandkids will fix it!

I used to have a pretty good size library which contained many old books from the late 1800s and early 1900s. They all had more knowledge of science than today's science books. We do not learn from history as humans; we repeat it. It seems as though birds have more brains than we do at times. If the nest gets attacked, they find a new location to live. Yet, humans build repeatedly on the water's edge and their houses get washed away. In Utah, people build right under the edge of the mountain, so the rocks can fall directly on their home.

Look at canaries in cages. When I was in school, the teacher said a canary kept in a cage indicated if the air was safe for the miners. If it died, get out! Now birds cannot survive the air quality in some cities, but people still live there?

The best fuel systems we build have been using old technology on the old dinosaur engines which have been around for 100 years. I have seen drawings and working prototypes of about 30 different types of engines which would make the old reciprocating piston engine obsolete. Sadly, people would rather go to a basketball game or a football game rather than helping any inventor improve their environment. It is sad to think people have been brainwashed into believing our government. All of these foundations and organizations are really saying and doing what the government allows them to say and do. Folks, this is just to keep the status quo.

Many times, people have told me, "If this technology really worked, then GM, Ford, Chrysler, Honda, etc. would all stand in line to use it." No!!! They have to comply with government rules and regulations. Others have said, "Just go to Greenpeace, Sierra Club, or any other organization and prove to them it works. They would love to use it to endorse you." Again, this sadly is not the case. These organizations can make more money ripping at your heart and wallet, talking about how horrible things are for these poor animals and our environment, and only your donation will help. Your donations go to new cars, jets, and good education for the children of the organization founders.

What the organizations mean to say is, "We do not want solutions. We want donations because solutions put us out of business." How many times have you seen the commercial for starving children asking you to support their organization? Think about it for a minute, folks. It takes a lot of money to buy TV ads, and the yachts and jets that these 501(c)(3) corporations own. Your money is going to support the organization, and many nonpaid volunteers will actually go out and feed a few people, or maybe a few thousand. The thing is, only a portion of the money ever gets to where it is supposed to go, either feeding some or cleaning the environment.

One of my original goals in 1983 was to build a product in America by Americans. They would be sold in this country, while giving back a larger percentage to the people in the form of jobs or equipment so they could also generate an income. Where possible, I want to help America first. I also felt every other country, whether it be in Africa or India, should have access to this technology and all other technology to put their people to work. It would help them build their own infrastructure and become independent.

Run It On Crude

Getting back to GEET, we found the inefficient injectors, when used with bubblers and other vaporizing systems, would work with different levels of efficiency. It took over four years just to develop one bubbler that worked for all types of crude oil, but it took 25 years to understand all the science behind it. In this book, we show and explain one picture at a time and describe how the different types of bubblers work in different discoveries. There is no one perfect package of GEET which can be mass-produced to fit every fuel-powered piece of equipment.

Back in 1983, I arranged for some meetings with investors in Beverly Hills. My stepfather told me I could use his backyard for the demonstration. It was my first GEET model that would run on crude oil which I brought down to demonstrate. First, I ran crude oil from Kern County, California and the test went very well.

There was a gentleman named Nick with two bodyguards who arrived in two separate Ferraris who asked if he could use his own crude oil. I said yes. He sent one of his bodyguards out to the Ferrari and got a bag, which was a plastic bag within a plastic bag. There was a gallon of crude oil in the inside bag. When Nick saw the look on my face, he explained he did not want crude oil all over the floorboard stinking up one of his cars, and we all laughed.

Once we tried his crude oil, he motioned for me to shut off the engine so we could talk. At this time, about eight people left the demonstration and went home. Many said the GEET system was out of their league. Nick said he was very impressed with the demonstration and had never seen any internal combustion engine that could run on unrefined crude oil. He also said he normally would not offer any money to any inventor until first his attorneys checked for patentable rights, checked the history of the inventor, and got a full business financial breakdown. I stopped him at that point and told him I was not interested in anyone taking over my invention. I was merely looking for some funding to get it to the next stage. With my comment, he started laughing and said, "You will never have enough money to get it to the next stage. The most the technology could be worth might be $5,000. It is obvious to me I am wasting my time talking to an ignorant inventor like you."

As he and his bodyguards walked off, I was shaking my head in disbelief. Now there was only one person left; I turned towards him, and I asked him if he still had any interest. This nice old man smiled and said he had only one request. He wanted my autograph. I looked at my stepfather with a quizzical look.

It was at this point a driver came up the driveway in a chauffeurs outfit and asked his boss if he could use the restroom. The gentleman turned towards me and asked where the restroom was. We all entered my father's home. While his chauffeur was in the bathroom, the old man turned towards me and told me his name was Hugh Johnson. I said, "You mean like Johnson & Johnson?" He said, "Yes, I'm the CEO." We shook hands.

I asked him how he wanted me to address his autograph. Hugh hesitated for a few seconds and began rambling on about how he felt his grandchildren would feel if they knew they had the autograph of someone as famous as Tesla. I turned a piece of paper towards him, handed him a pen, and asked him to write down his grandkids' names. I wrote each one an autographed paper. As I wrote out short little notes for each one of the grandchildren, I told him I felt very special by his request, but what I was seeking was financial help in getting GEET to humanity.

He was a jolly old guy, and boy could he laugh. After about a minute of laughing, he began to explain to me what he saw with GEET. It was his opinion if Johnson & Johnson sold off all of their assets it would not be enough money to properly market GEET. He said if every single piece of equipment which uses fuel on the planet today became obsolete, there was no company large enough in the world to replace it all. That is why he felt the best he could do was to ask for an autograph and leave me with a 2,000-dollar donation. I think of him from time to time and what he said. I realize that even Bill Gates might not have enough money to do such a task.

I think about how many of these little people have thought they could simply steal another man's invention and dreams and say they invented it so they could become the next Bill Gates. Bill, if you are reading this, I do not want to be the next Bill Gates. I am happy inventing and putting my mark on humankind as you have, but in a different way. Could it be possible someone like Bill could take a product like GEET and make it happen? Anything is possible, but who knows if it will happen.

Now if you coupled up a few of these great billionaires like Bill Gates, Mark Zuckerberg, etc., they might be able to collectively change the world to a clean planet. You need to throw in a few people who are successful in other corporations already and pull in the experts to come up with good basic designs as a starting point.

GEET Gas Is A Whole Different Ballgame

Well, how safe is GEET gas? This seems to be a very good question, but to answer it we need to look at how safe are propane or natural gas. Most people are not aware that both propane and natural gas have an odor added to them, or you would never know you had a gas leak. That is the same case for GEET. We do not supply GEET gas to any home or location where it could be a danger. If we did, we would probably make the GEET gas with a little bit of ammonia, vinegar, or something else to give it a slight odor that you would not like.

Since I do not like big bangs, I have chosen not to compress GEET gas. However, even with this warning, others have used propane cylinders and partially filled them under very low compression. I still do not recommend anyone trying this unless you are in a protected area and are more than qualified to be handling this type of an explosive. The danger level is way too high to take a risk. We have also found when we add a little bit of Old Spice aftershave, the exhaust smells like fresh roses, even if you are using up to 80 percent liquefied pig manure. When you tell people not to do something, it just makes some people want to do it more.

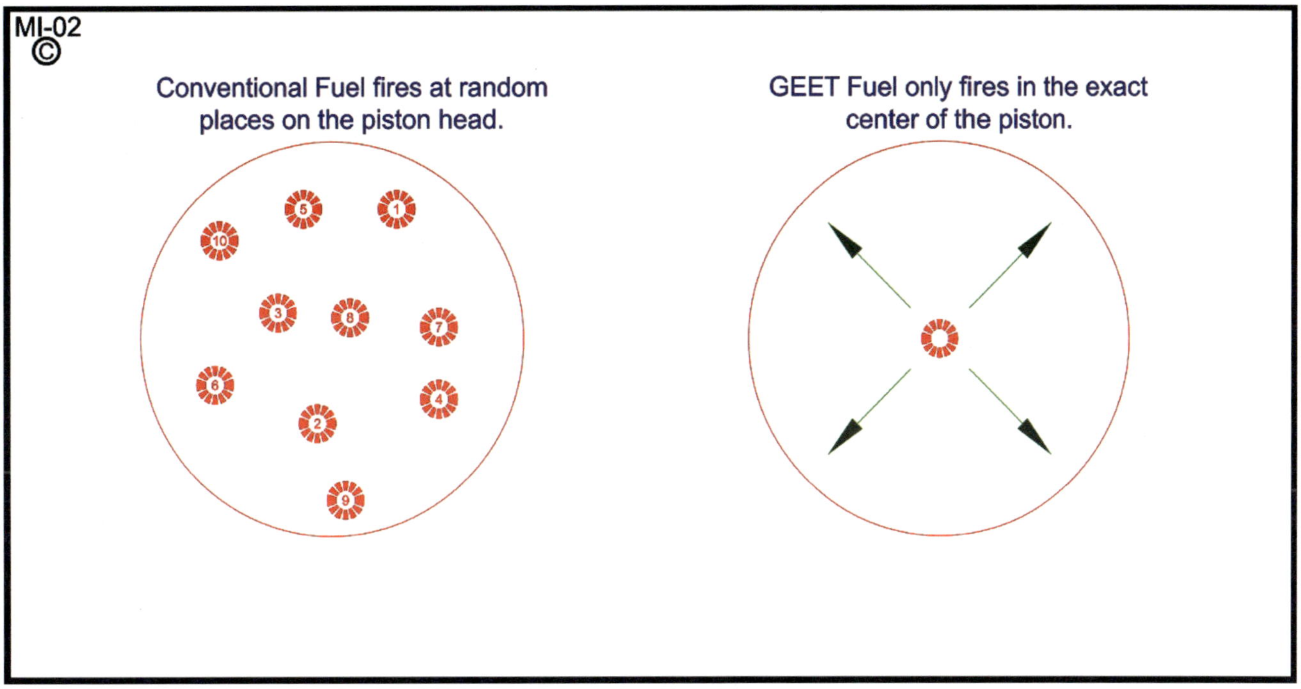

Some guys from Canada had contacted me back in 1994 at the Cambridge laboratories and asked if they could get one of my engines for testing. I sent them one, and they changed the head to a crystal head, so they could see what was going on inside the cylinder when GEET gas detonated. They phoned back a few weeks later and said they had never seen anything in the cylinder like this before. Every single stroke of the engine seemed to fire in the dead center of the piston. The glow moved outward to the edge and then back to the center. The lead scientist went on to explain normally only the vapor of gasoline or other fuels which went straight into a cylinder can burn, and the various sized unburned liquid particles coming out the tailpipe are what cause the pollution.

It was about two years later when the same scientist called me back saying he had made a mistake. He said when he looked at the camera footage on the same screen showing the position of the cylinder, the piston indicated an implosion had occurred as the piston was trying to compress the GEET gas. He further went on to explain by looking at the film in slow motion, the same footage showed the self-detonation process begin to occur before the spark came from the spark plug.

Well, thanks to our buddies at Cambridge, we can now explain how the engine would run cold since it had an implosion equal in force to the explosion. The implosion was pulling heat from the cylinder block into the chemical or electrical reaction. This seems quite logical. Since every stroke of the engine was in fact imploding and exploding from the center, the piston did not have the regular effect found in most engines. This explained why our 350 Chevy idled smoothly at 70 RPM. It is not rocket science, but I like it more.

You'll Get A Charge Out Of This

Let's go back to batteries. At the age of 13 in Quartz Hill, California, a farmer hired me to help on his ranch. On the first day of work, I got on the tractor and tried to start it. The old man I was working for and his wife were sitting on the front porch. He was in a body cast. He had fallen and broken his hip, which is why he hired me to get the almonds out of his trees before the first freeze.

When the tractor would not start, the old man hollered at me to pee a little in every cell of the battery. I looked over at him very surprised and asked him to repeat what he just said. He did. I turned my back towards him and his wife and did what he said. I then hit the starter and this time it started right up. It seems that warm uric acid in the urine will jolt a cold battery. The surge of power will allow you to start your vehicle quickly. A few students from Canada and Alaska have called me back to thank me for that one. It sure beats running around getting cables in another vehicle just to jumpstart something. This could be difficult if you are in a busy parking lot though.

Back in the late 1960s, I bought an antique BSA motorcycle. She was a beauty — 500 cc in a single piston, but it was a workhorse. When I first uncrated it from England, I could not help but notice one thing stood out above all. It was the battery — it was glass. You could see the plates, the acid, and all the buildup of little flakes of lead which were shorting out the bottom of the battery. The top of the battery was just taped on with black tape, so to clean the battery, you simply removed the tape, lifted out plates, removed the lead particles, and put the plates back in. You are probably already thinking it would be great to have a battery you could rebuild that would last you for the rest of your life. It sure beats the current battery technology which has you buying a battery that will only last for a few years and made so you cannot do your own cleanout or repairs.

I have looked at building my own batteries for many years, but realized there are too many government regulations on transporting batteries. It would make more sense to teach people how to build their own batteries which would last for a lifetime. If you are off the grid, you could make one set of batteries you clean periodically and never need to replace. Why have the battery companies hid this technology from the public all these years? The answer is clear. Money! If you build a simple lifetime battery, you do not need to buy more batteries. However, if they only last a few years at a time, you know the sucker, I mean customer, will be back every few years.

We are currently testing many different types of do-it-yourself batteries whereby we can just sell you the components. You can make great big batteries, assemble, and maintain them yourself. I have seen numerous types of homemade batteries, and it is very difficult to determine which ones you should build and which ones are just simply impractical.

Even with conventional lead acid batteries you can build a specialized tool to cut the top off your battery. If you are careful, you could clean the lead flakes out, put the top back on, add more acid and seal it to make it reusable for another one to seven years. On several occasions, I have seen universities announce they have discovered a new way of creating cheap energy in a battery. It usually comes out one to three years later some little inventor built one and put it on YouTube. The universities did not make these great discoveries, but they are always looking for inventors who are dumb enough to share their ideas openly and they are looking for big investors so they can take credit for the information.

If It Doesn't Work, How Come You Want It?

The system for advancement is usually done by universities claiming they made discoveries done by others. This way the government gives grant money to the university and keeps control over the findings. Many such discoveries, like the one from MIT resembling GEET, were sold to a company who put it on a shelf for a very large amount of money. Many people wrote to me about the MIT Plasmatron when it was released.

I spent most of the next week answering the emails. A few MIT students offered to stand behind me if I was willing to take down MIT for the theft. What was more surprising was the number of attorneys who said they would be happy to represent me if I wanted to go after MIT for patent infringement. I told them I chose not to at the time, and since it brought me so much publicity in a positive way and exposed MIT as being a bunch of thieves, I chose to leave it alone.

Is there a chance I would win a lawsuit against MIT, because many people know how corrupt universities are? No, probably not. However, all the attorneys involved would make a great deal of money and I would be supporting the 'legal' system. In most cases, I would rather dump my money down the toilet or hand it to an out-and-out bum, rather than handing it to an attorney.

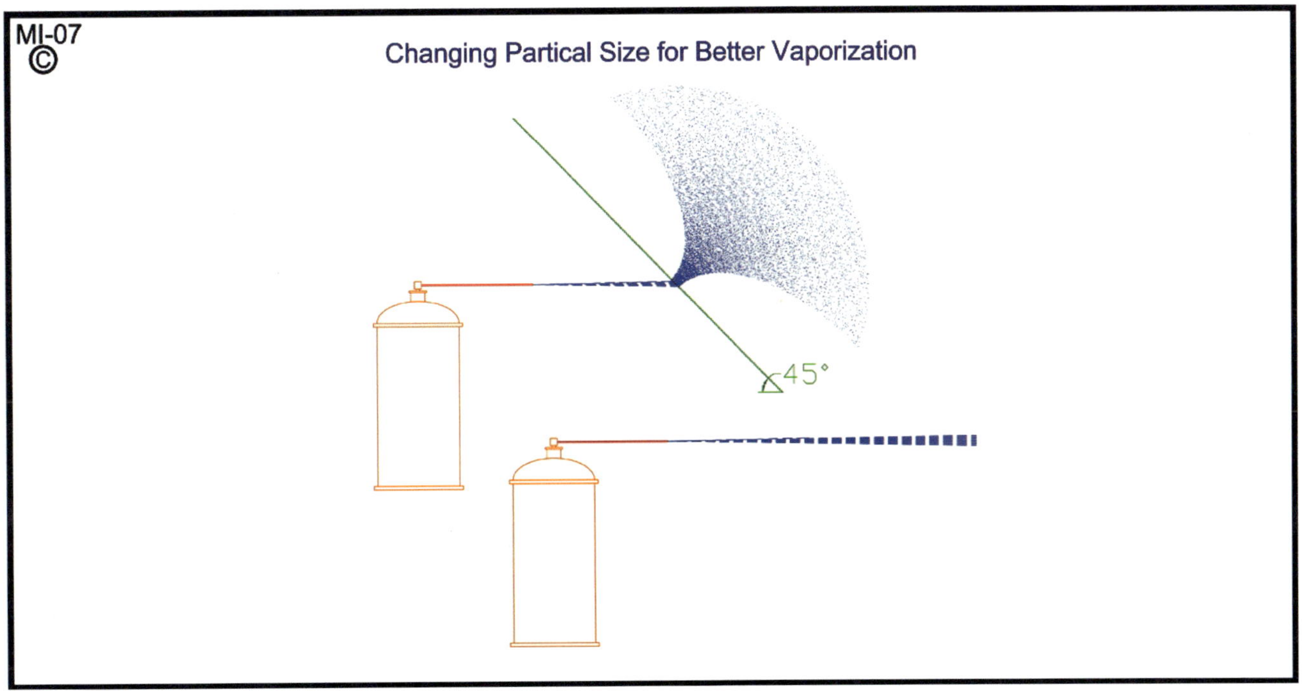

In the early 1980s, I was installing a fixed screen on as many cars and trucks as I could. Putting a wire cloth with about 40 fibers per inch made of brass or stainless steel at the correct angle (about 45 degrees to the stream of fuel hitting it) under a carburetor generated a 30 percent minimum increase in mileage. In some cases, we have heard people getting up to double their mileage from just this silly little thing, but not one patent.

When I did a patent search on "screens under carburetor," I became really upset with two patents issued in the early 1900s which obviously had been drawn by the same artist. There were about 100 patents for a screen to be used under carburetors, but as to why it worked, not one was correct.

During the tests in our shop, there were many times our gas analyzer indicated we had exceeded 10,000 RPM. It can be a lot of fun watching 300-pound men hit the concrete floor, covering their head, waiting for the engine to blow up. Their pre-programming had told them engines blow up when revved too high.

A New Element?

On the third day of testing at BYU, the three scientists who were doing the testing were very excited when their gas chromatograph indicated a new element with the atomic mass weight of 2.4. They explained this element is not on the atomic chart but it weighs more than a hydrogen atom and less than helium. As they were all laughing and joking about this new element, one of the scientists made the comment it should be called Pantoneium. I had to explain I had no interest in having my name on a new element which belongs to humanity. In France and Berlin, other scientists reported to me similar findings during their analysis.

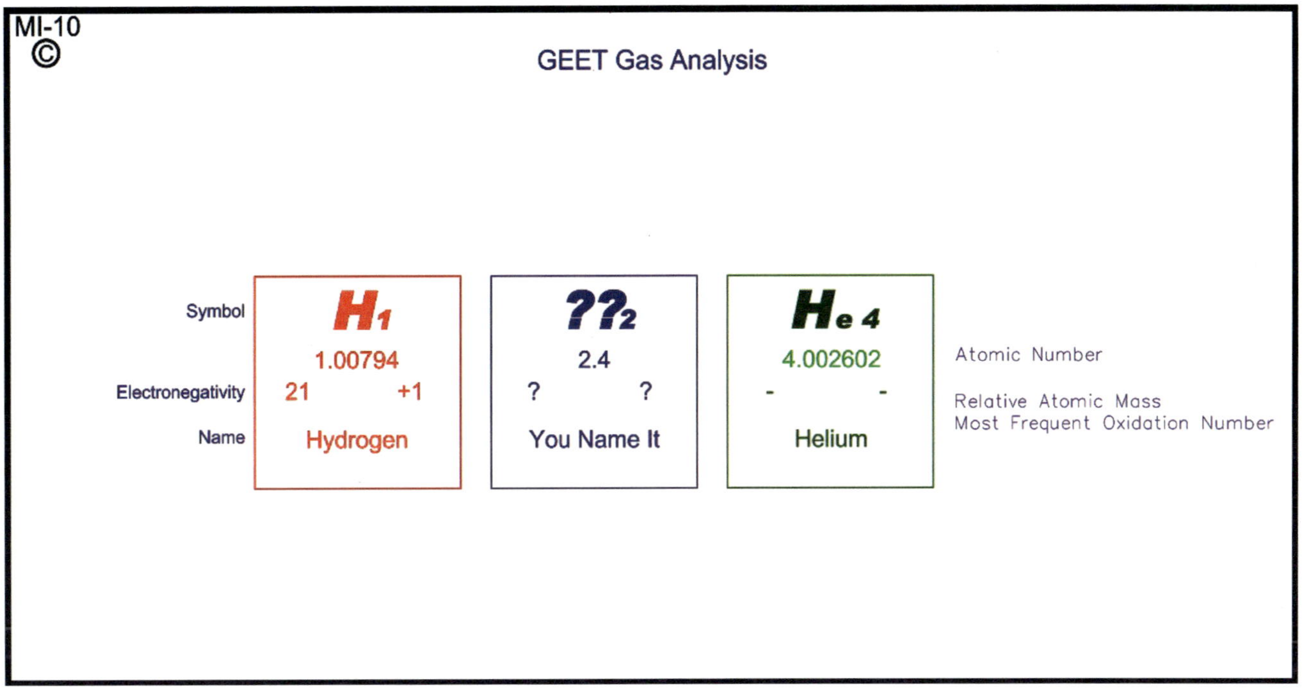

Why Is This Happening?

Sometimes it is the little things in life we do not pay attention to, so we shake them off as chatter. It is the simplest things in nature that can allow us to see what nature can do. Do we really need to spend all the money the government does? As you are reading through the following, why don't you make the same decision, as I give you examples? See if you can justify any reason taxpayer money should be used for these government-backed programs.

Example 1. Give anyone who applies an educational loan, so they can supposedly make more money when they graduate. Then you have government agencies arresting these people for not paying their educational loans back, while forgetting some other educational loans completely. This does not make a lot of sense.

Example 2. Let's say an organization receives five million dollars a month to study air quality in Las Vegas and Reno. We do not want them to actually solve any problems; we just want them to keep getting grant money every month to study the problem. I explained to the director of their program we could introduce them to the right people who could bring in weather machines so they would not need to fly around and test the air. By changing ionospheric conditions, one can bring a storm in or push it out. Tesla invented this device back in the early 1900s. He said the government did not want solutions; it wanted to study the problem. While standing there, many of his scientists said they could get in trouble if they ever had to admit they saw an engine with 20 to 21 percent oxygen coming out the exhaust pipe when there is not that much oxygen in Las Vegas.

Eisenhower wanted to build the best road system in the world. He wanted to do better than Hitler did with the Autobahn. A congressional meeting was set up, and a special group of congressmen was asked to review the findings to make a final determination. It took over a year, and the results were presented back to the special congressional meeting group. Three methods were highlighted and all were rejected because they worked.

One of these methods would use small particles of ground-up tires in place of sand and gravel when making asphalt. It would expand and contract but hold the road together with no potholes or cracks. Testing a second method, they took tar sand out of the ground, heated it up, and put it back on the ground with a heavy rolling vehicle to smooth it out. This did not take as much effort as putting conventional asphalt down, but the roads would last about 50 years with no potholes.

The reason we never heard about these two highway systems was because they competed with the system the oil companies wanted — one that would need to be repaired every year forever. Now, how could such a Senate or Congress subcommittee keep this knowledge from the president of the United States? Well, you need to look at where the money comes from with all the donations to their political campaigns, because this is where their loyalty lies.

MI-11
©

Orifice Configurations Can Change Air Flow Quantities

Air Flow Through an Orifice
V = CK√P
Q = AV
A = Area of orifice
C = Orifice coefficient
P = Pressure
Q = Volumn in cubic feet per minute
V = Velocity in feet per minute of .075 pounds per cubic foot
K = Constant based on air density of .075 pounds per cubic foot
K = 14, 786 when P = inches Hg
= 21,094 when P = PSI
= 4,005 when P = inches of water

EXAMPLE: CASE 2
V = CK √P
V = .92 x √21,094 X 0.1
V = 6137

Simulates an internal cone in the flow and provides a lot less pressure drop for the same amount of pressure and air density.

EXAMPLE: CASE 6
V = CK √P
V = .60 x √21,094 X 0.1
V = 4002

Simulates a washer in the flow and provided a lot more pressure drop for the amount of air flow.

Orifice Inlet Design From Fiat 500 USA Forum

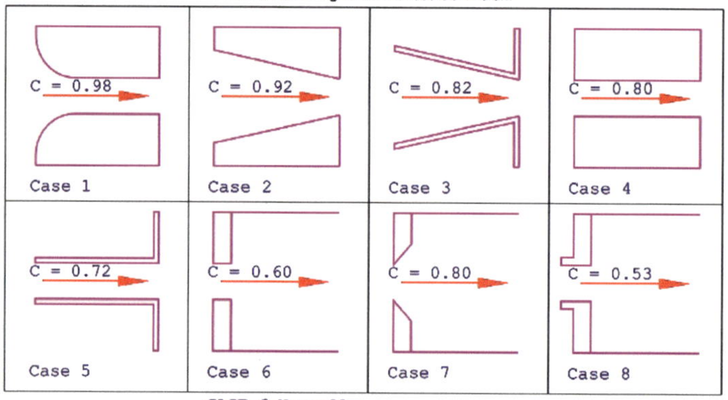

CASE 6 Normally used for all air intake points

We have mentioned already airflow is critical going into the GEET reactor, and we have one page showing all of the different configurations for airflow interrupters which would actually increase or decrease the efficiency. To comply with government regulations, we were forced to use the opening as shown in Case 6 to get the worst results we could. Using the wrong size rods, we could limit the efficiency to the 50 percent which the government allows. It becomes very costly when it only takes a few hours to build a GEET, but then it takes three months of detailing it to make it legal.

Get Grounded

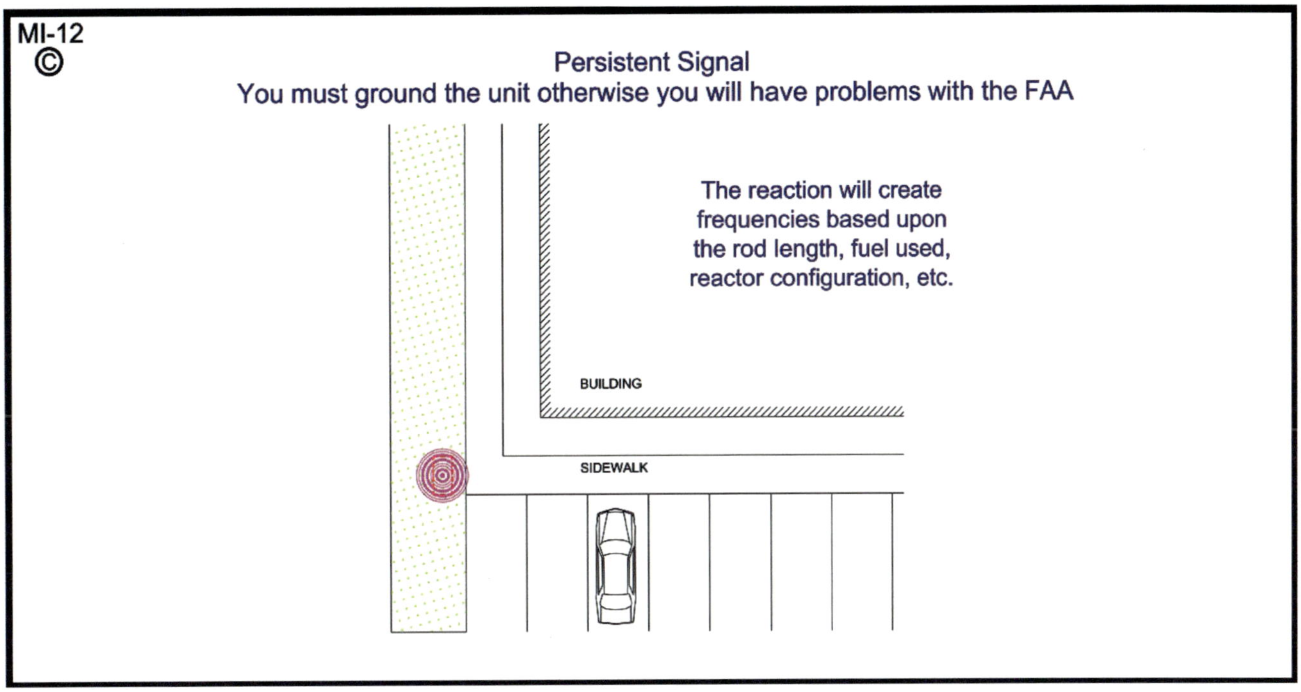

We have had times where people using a GEET system complained of black helicopters circling their homes, and so have I. I finally figured out why. If for any reason while GEET is running the grounding is broken, it becomes a transmitter, sending out a similar frequency to a downed aircraft. It was because of this we tell everyone to put a grounding strap at the south end of the unit back to the block of the engine to keep this transmission from occurring.

Do not use Teflon tape, pipe dope, or any other product besides Permatex Red RTV. To date, no other sealant has worked to stop the transmission of radio signals. Back in 1997, I described some of the dangers we discovered using pipe dope. I explained to my audience in Denver, Colorado pipe dope was the culprit behind all the steam explosions on the East Coast which killed or injured many people.

Things You Don't Know

MI-04
©

Thermal Inertia or Momentum - the Mpemba effect
(or Now you find out your Mother was right)

These uncovered containers were placed in a
refrigerator whose internal temperature was at 38ªF

39.5°F	at 25 minutes	32°F
38°F	at 25 hours	38°F

Back in 1979 in California and along the entire Pacific Coast, a breeze was coming in one evening and was dropping the temperature to about 38° F. This thermal mass hit all of the solar generating systems. Most of them were hot water back then, and the temperature change caused a low pressure which changed the freezing point of water. Billions of dollars' worth of solar equipment shattered. Since thermal inertia has never been written about or taught anywhere in the world, everybody was pointing the finger at someone else, saying it was defective equipment. I could have saved many of them millions of dollars or more, but no one would listen.

My favorite thing to do as a child was not playing with my friends. I preferred being around older people. Many times, I have been asked what I would change if I could go back in life. The number one answer was I would never have kept any job I did for all those years. I hated it. When I ask people why they stayed with the same job, they almost all said the same thing: "I didn't think I could do any better." The main problem with anyone advancing in any field or going into a new field is they are afraid to fail, so they do not try. The first 350 models of GEET I built were not failures; they were lessons of things I did not want to do again. It wasn't that they did not work, because they all did to one degree or another. I could give you 50 percent, 60 percent, or even 90 percent improvement in mileage, but I wanted zero emissions.

Many years ago, I was watching the History Channel special from the Howard Hughes foundation. It was about a satellite he sent up to take pictures while it went around Mars. Most of the pictures were not shared with our government. His reasoning was simple. It appeared there were three-, four-, and five-sided pyramids on Mars and other structures. A good businessman might want to go up there to find out why those things were there. Who built them? He was also aware it could have been a mining operation and possibly all equipment was left there. If so, he wanted it. At the beginning of this TV special, the foundation expressed that Mr. Hughes did not want this knowledge released until after his death.

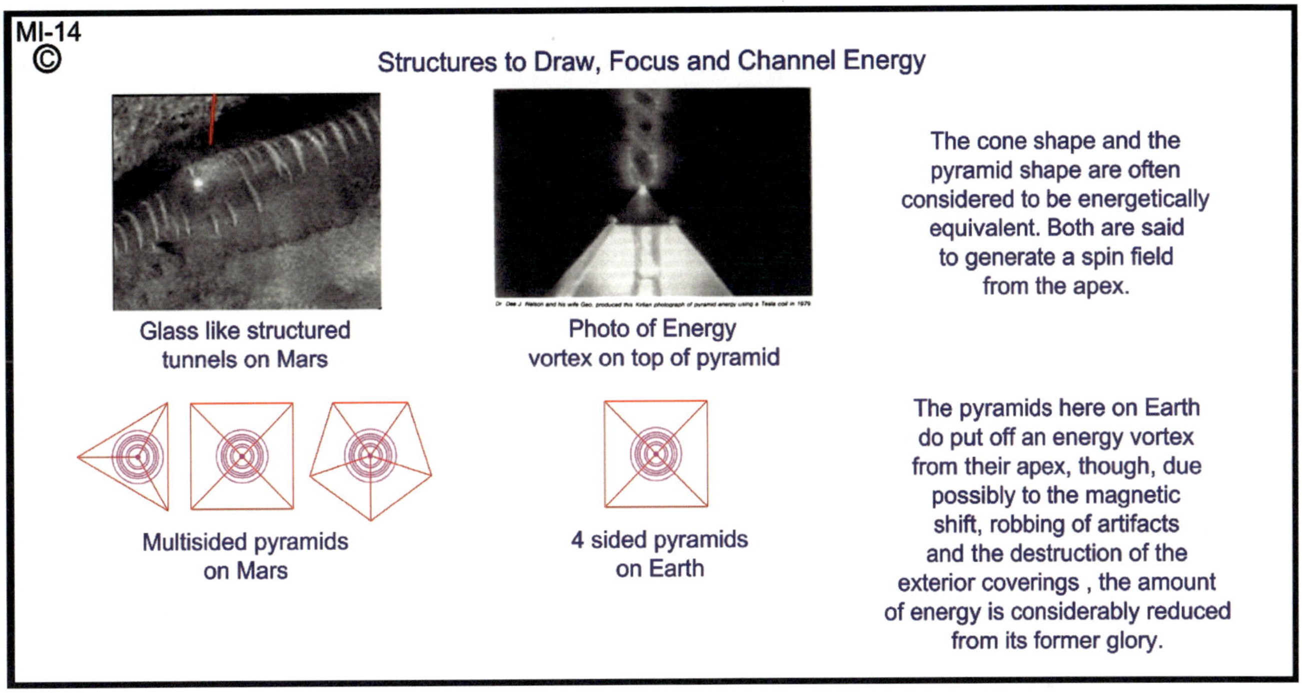

How many places are there on earth today which are seemingly older than we can credit for man being around? There have been many discoveries of cities and pyramids found under the water, and seaports under the Sahara sands. How many deserts or mountains have seashells at 3,000 feet elevation or higher? There are many anomalies, or anomalous findings on planet Earth, which have not been explained correctly. One cannot begin to understand how an item works without someone else's technological explanation. We have become so specialized, without a proper broad spectrum general education as a foundation, so we cannot see or coordinate with the other fields of science. How would you explain a timepiece found under the ocean encrusted with thousands of years' worth of coral and other growth, yet appear to be something from the future, based on the precision machining involved?

MI-13
©

Energy From Before?

Investigations into historical peoples have speculated they were able to tap into energy springs and constructing a grid pattern to harness them. Employing units of measurement based on simple fractions of the earth's dimensions, the ancients measured distances, in degrees, minutes and seconds of latitude and longitude.Soon a "prehistoric" network of dead-straight alignments appeared as if it sprung up everywhere at once invoking a large scale plan. Along these lines, (situated at terrestrial power points) arose temples, pillars, rocking stones, circles, crosses, mounds, pyramids, tunnels and platforms. Traces of these are dotted still throughout the world. You see them on every continent and even on remote islands. Sites had an exact geometric relationship to each other within a master worldwide pattern. In the words of John Michell, "a great scientific instrument lies sprawled over the entire surface of the globe." It is marked by these megaliths of all types, all aligned in a single geometric pattern--the remains of an ancient power network.

In order for it to operate, many sites were aligned to favor certain astronomical solar and atmospheric phenomena and terrestrial energy streams. They developed a technology to detect earth currents, and then to manipulate them to a predictable result. Thus, all over the world, thousands upon thousands of pillars (called "menhirs") arose in symmetrical formation, their purpose to realign earth magnetism from its natural meandering paths to straight artificial lines. At each pillar, the electric current of the atmosphere was attracted to combine with the terrestrial current to produce a fusion of power. (A secret we have not yet utilized.) Interestingly, the standing stones were all rich in quartz, a crystal similar to that used with the cat's whiskers in early radio receivers. The standing stones worked as cathodes. Researchers have detected an energy rising in spiral form from standing stones, gathering magnetic strength toward the top of the stone. Interestingly, photographs of the stones are sometimes marred by a mist of light surrounding their lower extremities.

MI-13 is a quote from "Dead Men's Secrets" by Jonathan Gray.

In 1875, a man had just closed his shop for the evening. He proceeded walking towards Times Square, where he stopped and purchased a newspaper from a young man sitting on a box. Two steps later, with his newspaper under his left arm, he seems to have vanished. Then at the same location in 1957, in the middle of the street at Times Square, a man appears and drops five or six inches onto the pavement. Rather startled, he looks up and sees this big yellow thing coming at him. He died on impact, according to the cab driver. His passenger said the man just appeared out of nowhere, with a newspaper folded in half under his left arm. Yes, the date on the newspaper was from 1875.

Hundreds of such stories come to the attention of newspapers and other reporters around the world with no logical explanation. The simple explanation is a time portal. Can it be proven? Yes, it can! Look at the Bermuda Triangle, or the Devil's Triangle, or the soon to be discovered Tasmanian Triangle. These intense electrical windows seem to open under the phase of a new moon. It is a shame the pilots and military did not know this back in the 1950s, when five Navy planes went off radar in the Bermuda Triangle and were lost. We do not know where they landed or crashed. If they went into the future, they may never have crashed or landed yet.

When Did Common Sense Die?

If one only studies architecture, he may get to know the field very well. As we advance, we seem to lose more and more common sense and only see the things we are told we should see. We ignore the rest. As an example, I have always been a carpenter; when I was in Seattle, an architect phoned and asked me to finish a building he was remodeling.

It was an old two-story church, and he had re-plastered the outside and then began on the inside. When I entered the upstairs, I panicked as I walked across the room and felt the building moving. It only took a few seconds to analyze what the problem was and how to fix it. I went to his office next door and explained the building was not safe and it was very close to collapsing. The last crew had followed his drawings and began cutting out all support for the roof system when they cut out most of the ceiling joists.

There were only four joists left to hold a huge roof and sidewalls in place. I drew him some sketches of how to fix it and he said "no!" I told him I would not go back into the building unless he had an engineer certify his drawings. When his engineer arrived and I took him upstairs, he only walked one-third of the building and returned quickly. He looked very scared. He said the building was not safe.

It took about three hours of the engineer arguing with the architect before he agreed to make changes. All of what I had suggested was accepted as being a safe way to complete the changes and still keep the basic concept of the design. Afterwards, the engineer offered me a job.

More Energy Than You Know

Over the last 37 years, hundreds of scientists and engineers have made the comment, energy released from the plasma within and around the exhaust of the GEET reactor most likely has more energy than the output shaft of the engine being fueled by the GEET system. It may take another 50 years to find out how to harness this power. We have done some preliminary tests and found the amount of energy at the south end of the reaction rod is normally ten times more than found to the north end of the rod.

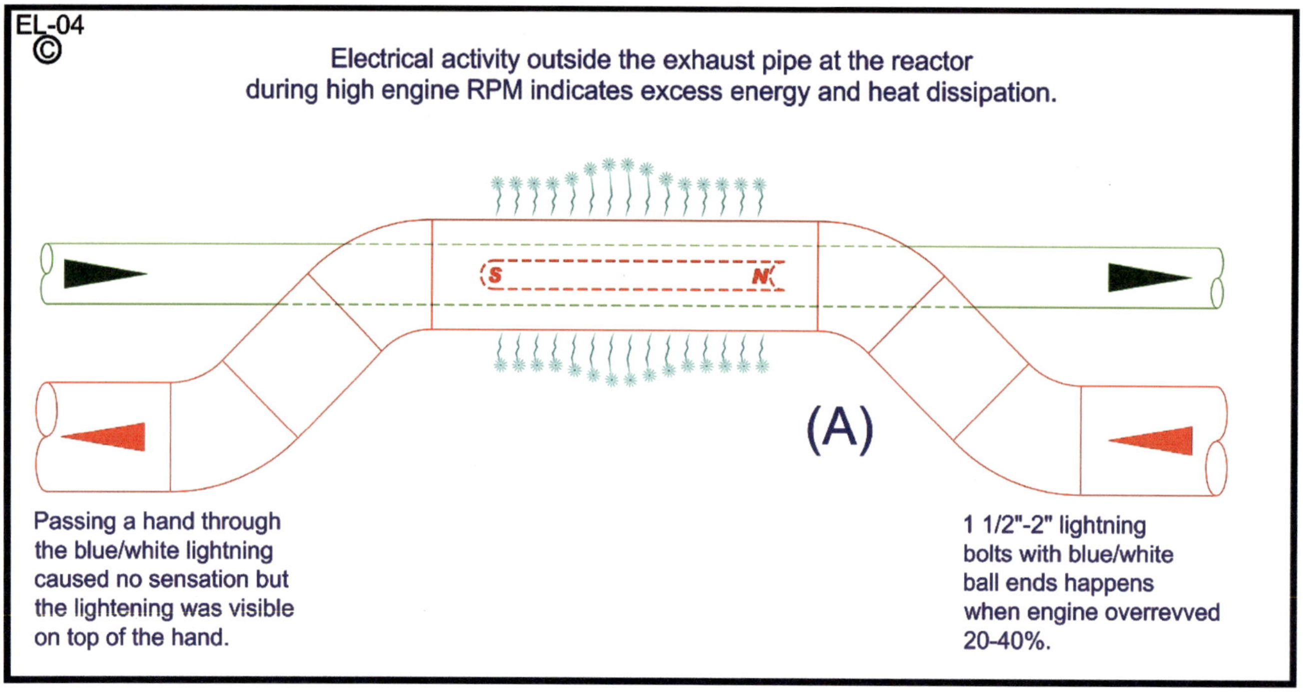

We do not recommend the following, but they are examples. These discoveries may help people in other fields.

If a ground is not properly hooked up between the exhaust and the engine or heat source, you could get one of three different types of responses and all are visible. The first one being beautiful little white and blue lightning bolts coming from around the reactor. Each bolt is about one and a half to two inches long with little blue tips. These small lightning bolts appear to be harmless and when you put your hand in their path, they will go through your hand and you will not feel anything.

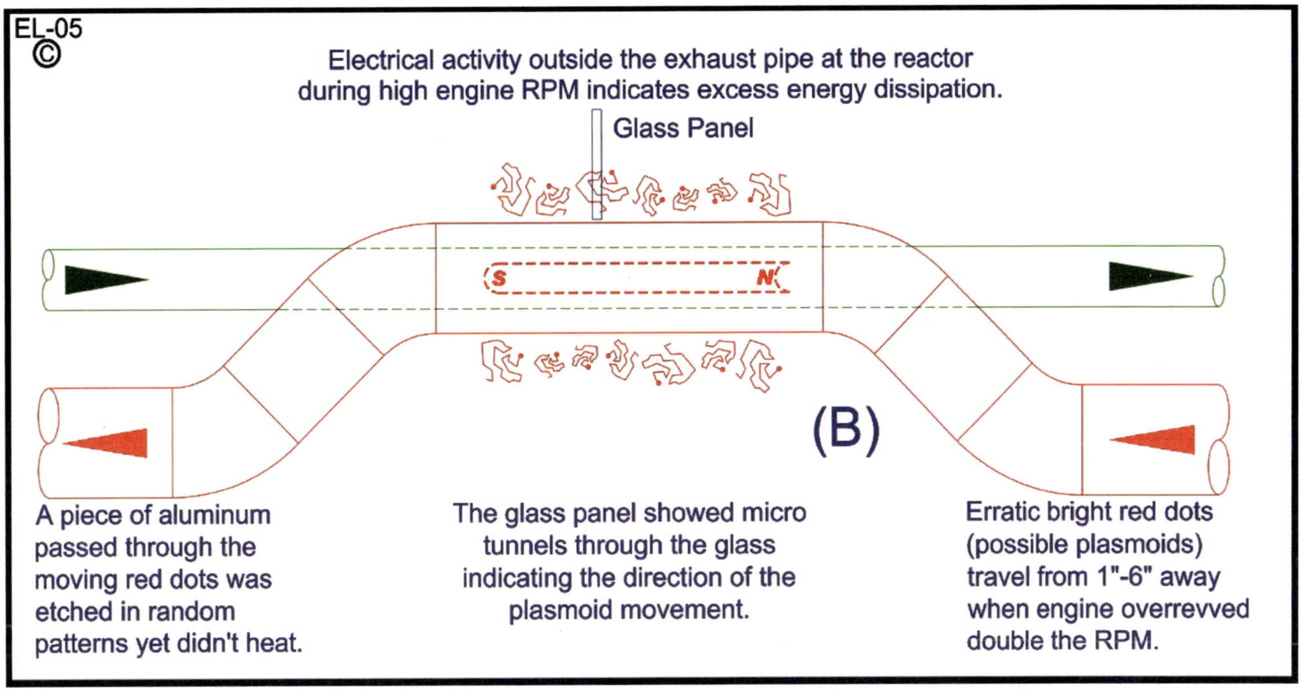

The second type of energy expelled from around the reactor is called plasmoids. We have found out the hard way plasmoids are very hot little devils with very erratic behavior. Let me tell you how we made this discovery, since most discoveries are made by accident.

We were shipping engines and generators out as fast as we could from Utah throughout the United States back in 1998. One of my technicians ran out of Red RTV gasket material, so he decided to substitute some pipe dope rather than driving thirty miles to the store. Once fully assembled, every engine was tested to make sure it was functioning properly before we drained out the oil and shipped the engine.

The technician put in eight ounces of fuel and started the engine with stopwatch activated. Just as he started the engine, someone turned on another piece of equipment which overloaded the electrical circuit in the shop and tripped the main breaker. All the lights went out. I ran for my office and the electric panel to turn the breaker back on. Just as I was reaching to turn the breaker back on, I could hear one of the employees hollering "Do not turn on the power."

Being a typical human being, my first thought was maybe somebody had gotten wrapped up in a piece of equipment or worse. I grabbed a flashlight and ran to the garage. When I entered, my technician asked me to turn off the light and look down. I was about two steps up, looking down at an engine running in the dark, and I could see bright red things surrounding the exhaust pipe of the GEET engine. As my eyes adjusted, I could see these erratic electrical things or pulses of light. They would move a quarter to a half-inch and turn. They moved in all directions, and there must have been 40 or 50 of them around the exhaust pipe.

I reached down with my right hand and put my index finger towards these little red dots; one went right through my index finger, coming through the fingernail and then directly back into my finger. I felt a jolt inside my finger. I let out a small scream and backed away quickly. While the engine was still running, I held a piece of eight-inch glass next to the exhaust pipe, allowing these red plasmoids to penetrate the glass to see if glass might stop what was happening. They seemed to be going through the glass without any interference.

I also had some aluminum flashing handy, so I grabbed a piece of it and held it next to the exhaust pipe to see a plasmoid go through aluminum. The aluminum didn't seem to slow them down at all. I told my technician to turn the lights back on. I wanted them all to get back to work. I returned to the shop area and I asked my technician what he did differently to this engine to cause such a reaction. He said he used pipe dope instead of Red RTV to save a 30-minute trip to town.

I sent my technician to the store to buy several tubes of Red RTV and I told him to build a new engine for the customer. I wanted him to stay away from this one, and we put ropes around it to stop anyone from touching it. I went to the house and spent about 30 minutes trying to stop the bleeding that had occurred when the plasmoid went through my finger. I then phoned Professor Eddie at a university who had told me he was studying plasmoid lights. Once I got him on the phone, I described what I saw with the GEET reactor. The professor said he was on his way.

The professor immediately came to my home in a borrowed plane to see if he could film these plasmoids for future studies. When he arrived, he began by setting up cameras and laptops and asked to be left alone. I tried telling him his cameras were too close to the reactor and he acted very insulted. He looked at me and very boldly stated he was a physicist and knew what he was doing. He went on to say I should stick with my simple-minded experiments. With this, I returned to my office.

A few minutes later I could hear cursing coming from the garage. About two minutes later, the professor came in with his head down and stated he was a little too close to the reaction chamber and it burned up his computers and cameras. He saw a bright red flash which first appeared on the lens of his cameras, and then the computer screens went blank.

Our professor was not totally out of luck. The light bulb turned on in my head, as I remembered we had a piece of glass and a piece of aluminum which might have some proof of plasmoids. I ran to the garage, got them, and returned, grabbing a big magnifying glass at the same time. At first, I could not see anything on the aluminum or on the glass to indicate plasmoids. The professor took the big magnifying glass and holding the piece of glass I had passed through the plasmoid field up to a bright light began laughing. He turned towards my wife and said he could not believe his eyes; he handed me the magnifying glass and the piece of window glass.

Only with the help of the magnifying glass could you see the tiny tunnels and little round rooms hollowed out into the glass. Most amazing to me was how one of the tunnels appeared to go halfway through the one-eighth inch piece of glass, do an abrupt 90° turn about one quarter of an inch, and make a return 90° turn to leave the glass. Similar discoveries were made on the aluminum. The professor said this was all the evidence he needed to prove the erratic behavior of plasmoids. Since he had also seen plasmoids for the first time here, he could tell his colleagues their theories were correct on this strange electrical phenomenon.

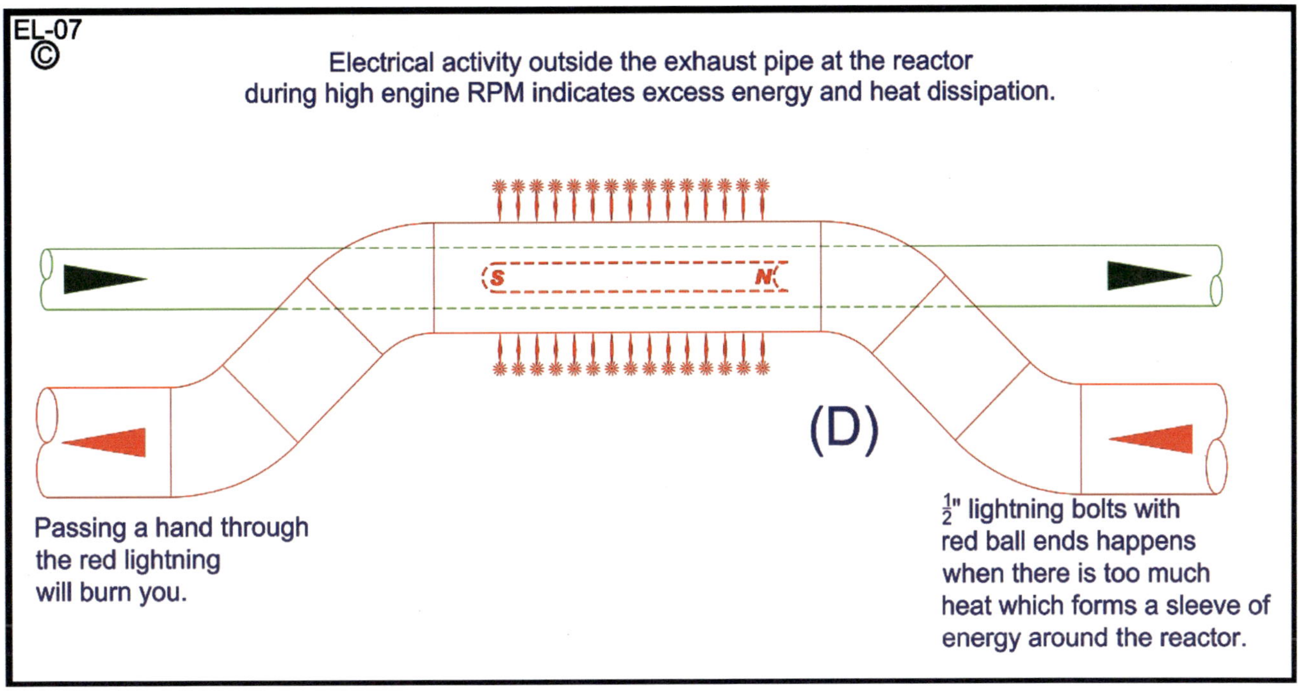

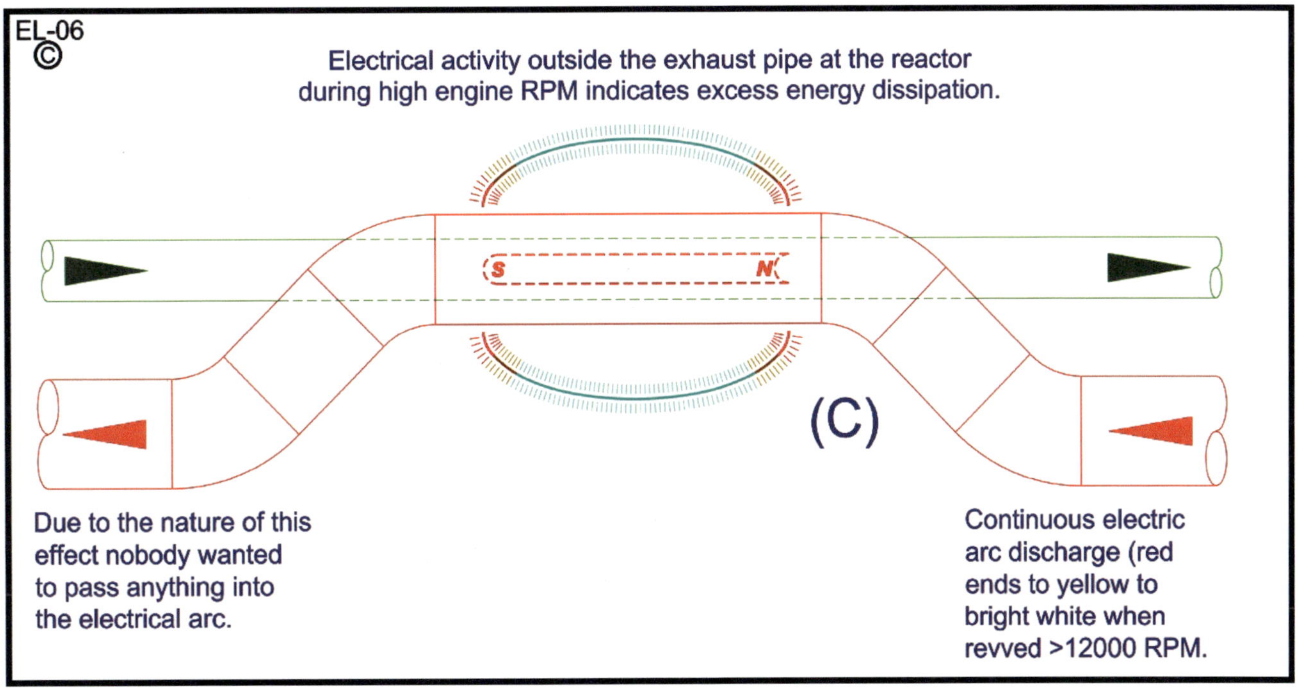

Last, we have total illumination as it was shown to Mr. Hormel (Hormel Foods) many years ago in Scottsdale, Arizona. Several scientists were invited to attend a presentation at his home in his private conference center. For this demonstration, I wanted to knock their socks off, so I put Teflon tape between the intake and the GEET, as well as around the exhaust pipe in the exhaust port of the engine. Just before the demonstration, I removed the ground clamp, leaving the GEET completely ungrounded in a horizontal position and pointed towards magnetic north. When the engine started up at an idle, you could not see anything.

I had told my assistant when I turned off the lights in the entire auditorium, he was to rev the engine up as high as it would go. The lights went off and the engine revved. Around the entire area where the reactor rod was inside the reactor chamber, there was a bright white glowing field. It looked like somebody had placed 1,000 watt light bulbs onto the stage where the reactor was and turned the switch on. For about one minute, you could hear the engine, but the audience was louder. They were all going "ooh" and "aah!"

I turned the lights back on and we shut the engine off. We did not want to attract any black helicopters. All of the scientists who were present made comments about having never seen anything like this before.

Over the years, we have explained this lightning has many different characteristics than what you see in science books. We have gone into details of lightning bolts, ball lightning, and all other types of electrical discharges we have experienced. Many of these types of lightning or electrical discharges can be replicated, and we want to find some means to photograph and chart them for future research around the world.

This research would have been completed if we had funds. We are hoping our situation changes soon as we put products out under license to other companies which will give us the freedom to do more research.

There have been several videos posted on YouTube where you can see electrical discharge occurring from the north end of the reactor going back to the block and surrounding metal. This indicates the south end of the rod was not grounded back to the engine and will most likely be transmitting a frequency. After working on one of our riding lawnmowers, one of our technicians used Teflon tape. He got about halfway across our five-acre field before he realized there were several black helicopters circling him. He jumped off the moving lawn mower and ran back to the house where we were all standing and laughing. Before he could say anything, I asked him if he had run out of RTV.

When he said he had used Teflon tape, we all laughed even harder. It does not matter how many times you tell someone not to build something without RTV, there is always one person who will not listen and will continue to do things their way. They always have.

Improve It Till It Doesn't Work Anymore

You are always going to have people who have tinkeritis. No matter how good they get something, they will take it apart and work on it until it does not work anymore. Sadly, most of them do not know how to put it back together correctly. When working with any invention or product, even taking apart a toaster, put the parts in the right disassemble sequence order, make notes or even make a video. We have had students of GEET who refused to take notes and were truly sorry later. One such student put it on his family van and took it for a test drive.

He did not return before class was over, so we shut down the classroom for the evening. When he did return, he disassembled everything he had made. He modified factory equipment and made more changes without notes. He went from getting three times the mileage to half his mileage. When asked what he had done differently, he shrugged his shoulders and said, "I don't remember."

We teach in class, "Don't modify any factory parts." I was so upset over his lack of cooperation I bought him a new carburetor, gave him gas money, and sent him home. No matter how many times we have told people "you must earn the knowledge by doing the basics first," people want to start with the most difficult project first.

Even today, as I write this book, I just received another email saying, "I don't wish to buy the full plans; I only wanted to buy the automotive section of them." Sadly, I will have to send him a second notice stating "We do not sell the automotive plans until someone has agreed to build an educational engine first." This is so they understand the basics of how GEET works. If you are a novice in a field such as electronics, would you expect a company to simply send you the schematics and all the educational material for you to build your own multifunctional cell phone with any possibility of you being able to build it? It is because of things like this I must refuse to sell plans to some people.

Patents, Attorneys And Inventors

We keep hitting all the different inventions. Some are not directly related to GEET, but many of our findings are within GEET discoveries as to how they work. The cyclone technology, which was patented out of Salt Lake City years ago, was a good concept if you are seeking a 30 to 50 percent improvement in mileage. But then again, you could accomplish the same increase by using a brass screen under an old-fashioned carburetor for a lot less money…if you still had a carburetor.

I was talking to an M.D. and as he explained different ways of treating certain specific ailments, I pointed out to him not everyone can use every product and get the same results. Some of us have allergies or a simple resistance to specific things. After giving him several examples, we both laughed but we agreed there is more than one way to start a fire.

Many discoveries, such as the first invention I patented, a simple log splitter, would be very difficult to market. However, another person who saw my invention decided to patent the same exact device as a door opener. He received a patent based on my design and operation of my device. Because he signed a nondisclosure and non-compete agreement, he is not calling his device a log splitter. Under the rules for a U.S .patent, as long as the claims are different, then you can use the same picture someone else did and get a new patent for a different field of application.

The sad part is so many inventions could be used in other fields besides the field into which they were patented. You would think your patent attorney would know this and would know which questions to ask so you could get a broad range patent, but this is not the case. Having hired many attorneys in my life, I have found they all claim to have expertise in their field, when sometimes they have no real experience in the area they are representing. One such example is an attorney in Provo, Utah who took about $20,000 in cash to represent me, claiming he had experience. He later told me he had never handled a case like this and did not have any knowledge in this field of law. Worse was when he stood up in front of the judge and said, "Your Honor, I don't work for Mr. Pantone. I work for you; I represent the court."

You know you are in trouble when the attorney says, "We can always file an appeal." This tells me he already feels he cannot win. Why go forward with him? I have done my legal homework before I hire any attorney, and I get upset when I ask for a motion to be filed and I am told by them it is not proper. I then show the attorney where it does apply in the law. Instead of being happy, they get really angry.

With experiences like this, you learn how to do your own paperwork and file your own papers, because you want to get them done right. Looking back, I wish I had known the difference between copyright and patent, but at this point, it does not matter. The number of discoveries and scientific breakthroughs which can be put in the form of literature are copyrightable. Now instead of wasting my hard-earned money to drag culprits into court for stealing, I need only contact the local federal authorities. They do their job, confiscate product, put people in prison, and it does not cost me a dime.

Before you begin sharing any idea, put it in writing. I would recommend a minimum of three or four copies and simply mail them back to yourself. I like to use my fingerprint as a seal on the envelope lip to show I did seal the envelope myself and the seal is unbroken. When I handed an envelope for each case to an attorney, the first thing these idiots did was open it up and destroy the evidence I had for a federal court judge. Having seen attorneys do this repeatedly, I have made many copies and sent them all back to myself by mail. This way when an attorney gets over rambunctious and opens one without thinking, I still have one the judge can open for himself. Then it can be admitted as a court document, otherwise it is third party or hearsay.

There are many organizations throughout the world claiming to help inventors do as I did. Take the time to call them. Ask them basic questions, like how many successful products have you made "X" money on or more? One such group in the United States advertises on television. They say they have had about one million people sign up for patent protection at 500 to 2,500 dollars per person, per invention to date. What I asked next was how many of those million people hired his firm to do a patent search and how much does that cost. He replied the cost for about 50,000 people ranged from 5,000 to 25,000 dollars. When I asked how many patents he had filed, he said only 1,000 or so. I did not ask him how much those cost, but I did ask how many of those products he helped patent or review ever made one million dollars or more. He was not aware of any of those people making any money from their inventions. Attorneys care about how much they make, not you.

You are either an inventor or you are not. Getting your invention to market is more difficult because, again, either you have the ability to market or you do not. Many people go on television or on YouTube or in newspapers advertising they are seeking money or backers for their product.

I am now assisting on a new "club" for inventors and people who are seeking to get their new products to market. This new clearinghouse will help put together people with a clear understanding of what each party expects from the other, prior to any agreements.

David Pantone, when 11 1/2 years old, rattled off 21 inventions in a period of 15 minutes. In one he described a Black & Decker waffle iron and told me, "On one side of the waffle iron it makes a 12" waffle for your lover and on the other side of the waffle iron it make 4 small waffles for your little loves like me." This was the last time I played "inventions with him", it cost $21.00

I would not expect an attorney to be out showing carpenters how to build homes or a brain surgeon teaching at a grocery store a better way to stock their shelves. So why would you expect an inventor to understand marketing? There is a big difference. Very few inventors will ever be able to give a proper sales presentation. Most cannot even explain why their invention really works, in proper scientific terms.

Just in the last few days, I have spoken with at least five new inventors I had never talked to before. Each one had his own definition of what he was doing, and I let them explain their process to me. Once I understand them, I apply the knowledge I have to show them maybe they missed the boat on a portion of what they had invented, because they were not looking in the right place.

About 20 years ago, an inventor was showing me his new way of pumping water further than anyone else was. His goal was to patent his new pump. When I saw his new pump working, I tried to explain it had more value as a desalinization unit than it did as a pump. This inventor became so upset with me, he called me names and told me I was fired. He did not want my help. He patented a method of pumping water further than anyone else has ever done with the same basic pump. So many times people patent things that could have been much better had they had a broader knowledge and vision of other potentials for their inventions.

Many inventors who come to me seeking advice get the same advice as investors. I tell people who want to invest in me I am the worst investment in the world. The best investment is investing in what you know. Do not invest in someone else. I give inventors all of the pros and cons and I give them free advice. I do not ask money for helping a person who has already spent all of his money building their invention. I explain a better way to make their invention more efficient. If down the road they choose to donate some money to help, that is fine, but not mandatory.

Give Credit Where Credit Is Due

I also feel what is mine, I have earned. I expect a royalty or payment for using the GEET in any form. Thirty-five years ago I mentioned I would be happy with just a dollar a car and fifty cents per bicycle or motorcycle which are motorized using a GEET. There were no takers. I have tried also to convince governments by adding water to fuel we could increase mileage significantly. Again I had no takers.

We all know running a car on water would definitely break the bank as far as governments are concerned. This is why in 1983 I suggested to people in government they create an odometer tax. The government could compute how much mileage each car should get with basic factory equipment. You can calculate it might be one cent or two cents a mile for every mile on your odometer. For high mileage people, they might even have to do that twice a year, but this way it would pay for all the road repairs needed throughout the country.

It all seemed logical to me, but then again, I am not a politician. Most of them said to change existing paths of income to new ones is very difficult. The oil companies would still make all their money off asphalt and other products, and by using the same GEET technology, they could turn the gasoline into a good clean fuel to run power plants.

Complain About It Or Do Something

In Utah, Kentucky, and places where they have coal, people complain about the pollution. We again are using Stone Age technology. If we used GEET, we would have zero pollution coming from coal. We have even run the small GEET engines on coal dust. We mix it with water with a small amount of soap to deliver the coal to the cylinder for a full clean combustion.

No matter how many demonstrations we have given, people who make major decisions have too many people below them who have a vested interest, or wish to have job security. Of course, an engineer working for Caterpillar or Cummins is not about to go back to the company and say, "Look what we found! It is so easy to build even a 10-year-old or 12-year-old kid can build it." For them to do that would be suicide. Who needs the engineer now? So instead they go back and say it is years from being possible or practical.

Here in the states we had a company called Waukesha who built good generators and pumps. Back in the 1990s when their engineers came out to see a Waukesha running with zero pollution, they reported back to corporate they had not seen anything that worked. Waukesha spends millions of dollars a year in R&D trying to get a two percent reduction in pollution. Years later, unable to compete, they went bankrupt, only to be now owned by GE.

In Chicago a few years ago, a trucker bought the plans and said he was putting the GEET system on his truck. I assumed, since I told him it was only for personal use, he was putting it on a pickup truck and not a large commercial truck. I gave him the details of how to hook it up. Later he called to tell me he was getting triple the mileage on his Cummins engine and the IRS caught him. I asked, "Why would the IRS be interested in someone getting better mileage on their pickup truck?" This is when he told me it was an over-the-road semi. When he was being audited, they noticed a discrepancy for the amount of fuel he had claimed he purchased.

When he arrived for the audit, his fuel receipts would only justify one-third of the miles he drove that year based on six miles per gallon the engine should get. He smiled and boasted he was getting over 18 miles to the gallon and zero pollution. They said, "It is illegal to modify commercial equipment without a permit or license," which he did not have. He still owed all of the sales tax and road tax for all the gallons he did not buy. This poor guy went from feeling like a proud American to an embarrassed thief, stealing money from his government. When told he owed more than $50,000 plus penalties, he decided it made more sense to give them the keys and quit trucking. At that point, he felt maybe he should just go home, go on welfare, and make more money than working.

How many other great inventors have there been or will there be who could really clean up this planet in short order? I do not know. I am looking for all of the special inventors to see if we can get some of these products out now. It does not need to be five times more efficient. Let's just give the proper political correctness the government wants, as it still has a positive impact.

See Through The Subterfuge

Maybe all of us know some things we learned were just flat out bull crap. When you know this, you can prove it. That is when things get to be fun. In 1983, when I first had my engines running on what I call fun fuel, I would brag about it. Some overeducated fools throw money on the counter and bet me it could not work, and I always covered the bets, to my profit.

One very wealthy investor, after watching the engine running on crude oil and soda pop, asked me to pee in the fuel tank. Back in 1983, I did not know if it would run on urine, and I said so. I took the lid off the fuel tank and peed in the tank. When I pulled the rope, I did not expect the engine to start right up, but it did. We both laughed until the engine stopped about ten minutes later. At that time, I said it must be the beer I just recycled. We laughed harder. He really did give me a check for 20,000 dollars, and on the bottom, he wrote "gift."

From then on, I could not wait to see what else it would run on. People were bringing me all kinds of junk cleaning solvents from body shops and transmission fluid. I tried everything. It did not matter what I tried. It all worked. My suggestion to you is if I have not talked about it, you might want to try it. It just might work. Let me know. I would like to add to the list of what is good fuel.

We will keep trying to find our niche to get the products out without interference. I made one special engine for the U.S. Forest Service to run on paraffin blocks. This way they did not have to have so much stored up fuel in the forest. Before they could use it, they had to get special approvals, and they asked for EPA approvals. When I inquired with EPA if I could obtain an approval for a generator or engine to be used only with paraffin on U.S. property, I was told I had to volunteer for testing all equipment the technology would fit.

I have talked with many inventors who have volunteered to work with EPA, and none of them ever received final approval. I declined to volunteer for any of their testing. I remember Dr. Wood telling me that every time he passed the test, there seemed to be just one or two more to be done. Every year it seemed like they had to reschedule lab time and were told to come back later. Using government-approved laboratories is very expensive.

But then again, we have a government who will spend 10,000 dollars on a toilet seat or hammer. If these money wasters were held responsible for spending taxpayer money, they would have to reimburse the taxpayers and face prosecution. Waste would be stopped or drastically minimized. Building a product in the United States has become very difficult, because of all the trade treaties we have with foreign countries. Yet, we do not see lower prices when we spend our money at the store. Who is getting all this extra profit? Is it just big corporations, or is it the government or a combination?

Years ago a teacher at a local Utah community college wanted to train prisoners at the jail facility to make GEET units. The program went for a very short time before the warden told the teacher, "You are not allowed to teach our prisoners a skill. We want them to be repeat offenders." The teacher was then asked to leave the prison.

I know it sure is not us when I hear a shirt can be produced for as little as 25 or 30 cents including all the labor. Why do we still spend 15 to 25 dollars for the shirts? There are too many people in the middle taking double and triple markups as it passes through their fingers.

Can you imagine pulling up to a gas station with this sign: "This is an environmentally friendly service station." To begin with, there is a big generator behind the building producing two or three times more electricity than can be used on the property, and the excess electricity cannot even be given to the utility company because it would offset their hard costs they need to verify to raise the rates next year. The fuel power for the generator comes from the vapor off the underground storage tanks which can produce it free, instead of the vapor bleeding into the atmosphere. Yet no one is willing to do it.

Let's just look at electricity first. Why is it so expensive? Well we have to pay for all of the people who built the lines, put the power poles in, built the dams, and for maintenance and retirement benefits. This is just for openers. Yes, I will admit I have seen some very hard-working people with utility companies, but on the other hand, I have seen tremendous waste.

Many years ago, company loyalty was extremely important. Now the companies want to make all the money and pay you minimum wage. They do not want to guarantee any kind of long-term benefits for you and your family. These savings means shareholders have much more profit.

When you use nature to create energy movement, it can be done on a very small scale, or a very grandiose scale. The solar heating tower, as can be seen on the California/Nevada border south of Las Vegas, are used to make steam to make electricity by using gigantic reflector arrays. It is the same basic principle as using a magnifying glass to set a leaf on fire. The engineers involved make it seem like it is a big new breakthrough! There have been very few breakthroughs in energy the last hundred years, except in the electronic field.

The big car companies are using electronics and computers as leverage today to convince people they should not be allowed to work on their own car. To have car companies come out and basically say they own your car and you do not have the right to repair it leaves a very bitter taste in my mouth.

Push And Pull The Snow

If the vapor coming from the bubbler on the GEET system can be pulled off in the direction of spin, the efficiency of the burn will be increased. Once the vapor reaches the reactor rod and is still being pulled by the same amount of vacuum, but into a very restricted flow area, it is gaining temperature at a rapid rate. This is very similar to the effect of a black hole, where matter is ripped apart to its lowest possible atomic weight, even pulling all light into itself. Light is heat. Certain heat frequencies have been used in the treating of many diseases and will become a more common practice in the next 10 years, if not banned by WHO or other medical associations.

Nikola Tesla built a weather machine which would spin, creating a high-pressure and low-pressure zone on both sides of his device. The low-pressure side could be aimed at a cloud bank, pulling the clouds into you. This would allow a large amount of rainfall or snow to be brought to your specific location. The high-pressure zone could be used to directionally control clouds coming at you and divert them to a different direction away from you. This would affect about a 120-mile diameter of weather.

Although this might be needed in some areas, the results can be devastating if improperly used. In the 1930s, the notes of this invention were appropriated and now the USA has HAARP, and other countries are using similar technology. Although it may seem like a great idea, it is, but only in the right hands. To simply create floods or droughts for making money, in my opinion, is very sick. On a grander scale, any country who can control the weather can control the food, water, and highway systems. As a side note, our planet is an integrated system and if you take moisture/weather from one place, you can expect a corresponding problem to appear in another place.

While in possession of one of these weather devices when in Utah, just before the Olympics, I had a brainstorm. I contacted the Olympic Committee and asked if they might be interested in having it snow from midnight until 6 a.m., and then beautiful skies from 6 a.m. to midnight? They said they would get back to me in a few days, and they did, only not directly. The call came from the governor's office. I was told if I turned the weather machine on to do what I had planned, I would be held responsible if anyone was injured or any property damage occurred during this time.

It was clear using the weather machine, even on a local scale, could cause danger. I returned the unit to where I received it and thanked them for considering me to use it. The only reason I returned the device is that I have been ripped off so many times by government agencies, employees, and unknown others I did not want this technology in someone else's hand.

No Thank You

Back in about 1998, I received a call asking if I would be interested in being nominated for the Nobel Prize. I did not hesitate. I said no thank you, as it is no longer the honorable group it used to be. Too many people who are not worthy of the prize have received the honor, and therefore I do not wish to be part of an organization who will honor this kind of person.

I value the emails and letters I received from individuals such as yourself much more than having a PhD diploma hang on the wall or a Nobel Prize on the mantle. I do not look at titles as distinctions of honor, because most people who have titles really have not earned them.

We have had a few times where the our normal tests done on GEET did not indicate a plasma reaction was occurring, yet the outcome of the test indicated it was working. If solar flares are interfering with the conditions between the earth and the sun, it will stop the compass from showing a healthy aura or auric field around the rod. Once the solar flare has been completed and passes the earth, the rod goes back to its normal auric field. There is a lot more happening here than meets the eye and this is another field needing more testing.

Real School Of Higher Education

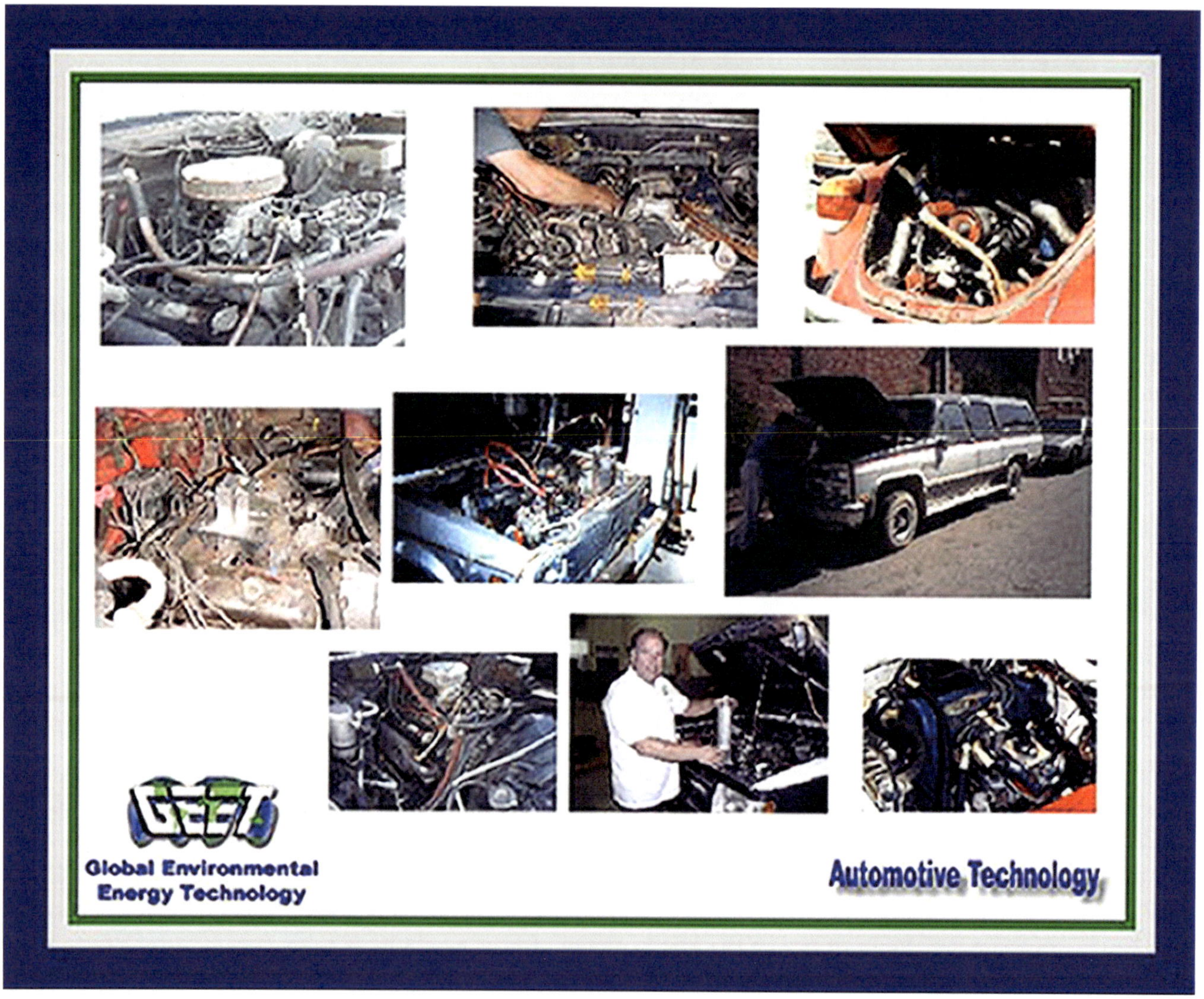

Back in 2001, I had a visit from a gentleman representing the Utah State Board of Regents. He said other universities had filed complaints claiming I did not secure the proper permits for my school. I escorted Mr. Carpenter into the classroom, where the students were building their first GEET demonstration engines. When Mr. Carpenter looked around the room, he could not help but notice all the nametags. At that time he turned to me and said, "There are five people standing at this table, and four of them have PhDs in physics?"

I smiled and told him they were my students that week. Pointing to the young man at the end of the table, I explained this ninth-grade dropout, me, was teaching them how to build GEET.

Several of the scientists there had already built the first engine, and their excitement level was rather high. The conversation between the scientists and Mr. Carpenter lasted about 20 minutes. A few of the scientists told him I did not charge enough for the class, and the 2,500 dollars tuition was a pittance of what the class is really worth. One scientist went on to explain he learned more from Paul Pantone in one day than he learned in college in 12 years!

Mr. Carpenter told me to follow him to his office and get my license to be recognized as a school of higher education. He knew many people would be upset at the universities, but how many of them could say 25 percent of their students hold PhDs. I do proudly carry that license today in my wallet. Shortly after the Utah State Board of Regents issued me my license. Now, the director of the Board of Regents can no longer issue licenses for universities. The Utah Attorney General's office handles this new task.

How To Make It Work

It is funny how many ex-employees and others have claimed they really are the inventors of GEET. On one occasion, I went to a small engine shop to place an order for some carburetors I needed and the owner of the shop asked me if I was working with the inventor, Mike? I was with my brother that day and we both had a puzzled look on our face, but I proceeded to explain to this gentleman I did not have an employee or had one at that time named Mike. My brother asked if the unit he had seen running the GEET technology was on a red lawn mower and was told yes. It seems as though our employee, after modifying my brother's lawn mower, claimed he was the inventor. I showed the owner the magazine articles and handed him my business card. We went ahead and ordered a dozen carburetors, not knowing they would never arrive.

About a week later, there was a horrible earthquake and tsunami in Japan. My carburetors, along with millions or billions of dollars and many lives, are sitting at the bottom of the Pacific Ocean. It appeared at this point we had to come up with our own fuel supply system and did not buy parts from anyone. We may have to set up our own foundries and systems to make each and every part ourselves, so other governments, shippers or natural disasters would not be able to control our distribution again.

When using GEET gas to run furnaces, boilers, stoves, etc., you need to reduce the orifice size to about half the size of what you use for propane. You also have to redirect the direction of the flame to allow for boundary layer heat distribution. If the flame hits the heating surface at 90°, it will burn through the surface very quickly, but at a 45° angle, it will pull in a boundary layer of fresh air between the heat source and the material being heated, mitigating damage.

Aerodynamics within a partial vacuum have different characteristics than the same vapor under pressure or ambient conditions. Now some people with school training begin to get GEET quicker if they have had a heating and air-conditioning background. Mechanics on the other hand have tremendous difficulty grasping the very basic parts of GEET. Since engines are designed to waste a tremendous amount of wet fuel, the formulas mechanics learn only apply to the conventional carbureted or fuel-injected systems. We have even tried propane carburetors and natural gas carburetors, but those systems were designed to work under pressure, not vacuums. Our most successful operation used a propane carburetor and brought the cost up to almost 3,000 dollars to convert a small car. However, using a standard small lawn mower carburetor from an 8 HP engine on the same car got the cost down to about 250 dollars for all the parts to build a GEET.

We have tried using the old carburetors that came with vehicles, only there are so many air leaks the vacuum cannot be maintained at the seven to eleven inches of mercury needed. One Jeep was using a radio-controlled carburetor from a small model plane, and it was still overkill.

A system of modifying vapor needs to be invented for proper delivery of fuels to the reactor for maximum efficiency. The control mechanism for this may be vacuum or based on the RPM of the engine. We are currently using ultrasonic diffuser units to bring the particular size down to approximately one µm. In different parts of the world, members of our R&D team are trying to achieve the same goal using different methods.

They Can't Stop It From Working

In the early 1930s, a gentleman named Charles Pogue became very popular for his carburetor. There are many books written on this carburetor and many articles on the Internet claim that he was getting 200 miles a gallon. Some people have a big imagination, but it was enough for the oil companies to be worried about. They had Congress use taxpayer money to set up the Fuel Research Institute in Tulsa, Oklahoma. This new group had a big job. Their main goal was to stop Pogue's carburetor from working. The oil company supplied scientists and assistance to change the formula of gasoline so the gasoline could not be properly vaporized. It would show his unit as being a total waste of time.

Once they knew they could change the formula and other cars on the road would still run, the military released Mr. Pogue from any military obligation and wished him luck. He spent all the money he had earned from the U.S. military and borrowed more to go into full production. He set up and trained dealers all over the United States and Canada.

About six months later, he began the distribution to all of his licensed dealers. Soon there were Pogue carburetors coast to coast, from the Atlantic to the Pacific, and everything was going fine for about a month. Then the new formula of gasoline was introduced to all the gas stations, and it did not take long before all the extra lead and lacquer would settle to the bottom of every Pogue carburetor, causing them to fail.

Pogue was beside himself about being forced out of business. Sales had been so good, he took investor money for expansion and now everyone one of them wanted their money back: the customers, dealers, and investors. In 1999, I had the pleasure of talking to Charles Pogue's great-grandson, David Pogue. When I asked about his great-grandfather, David told me he had just donated all his paperwork to the University of British Columbia in Vancouver. He informed me the carburetor never got more than 46 miles to the gallon, and all of the stories on the Internet were made up.

About three weeks later in Portland, Oregon, while I was attending a conference, I met Joan Pogue. Joan is a great-granddaughter of Charles Pogue. She began to tell me all about the Pogue carburetor, and I stopped her when she said it got over 200 miles to the gallon. I told her David Pogue had told me it never got above 46 miles per gallon. She told me she never heard of David Pogue. I explained he was the great-grandson of Charles Pogue, and she stated that great-grandfather did not have a great-grandson named David.

I went on to explain that he had just donated all of Charles Pogue's designs to the university in Vancouver. She said, "That is funny, because I live in Vancouver, and I have never heard of him."

When I pulled out a card that had his name and phone number written on the back, she said she did not recognize the area code. I informed her it was the area code for Vancouver, British Columbia. She began to laugh and explained she lived in Vancouver, Washington.

I then called David's cell number and told David I wanted to put Joan Pogue on the phone to speak to him, because she was the great-granddaughter of Charles Pogue. He laughed and said sure. I handed the phone to Joan and walked away laughing. How many times have stories changed so much even the relatives do not know the real facts?

It gives me great pleasure to say that in 1999 when I got to Oklahoma City, I contacted everyone in my directory for 100 miles around to invite them to come see me and see the GEET engine run first-hand. A young man who answered one of the numbers I called said the person I was trying to call had left his cell phone in the back of the truck. This young man offered to take a message. Once I identified myself as Paul Pantone, the inventor of GEET, he went off on me. He said it was my fault he was losing his job and hundreds would be unemployed because of me.

When I asked him what he was referring to, he informed me I had just called the director's phone number for the Fuels Research Institute. He also said they had been spending years with hundreds of GEET engines in the laboratory trying to find anything that would stop the GEET fitted engine from running, and they could not. Since they could not sabotage me by changing the formula of gas, there was no need to keep the Fuels Research Institute going. Everyone was given a slip to go home.

Had Charles Pogue designed the GEET instead of what he designed, we might all be breathing clean air today. Or would we? Almost everyone I know has said they would have taken the offer from OPEC and sold GEET and all the technology for a few billion dollars and just said to hell with mankind. I could not sell out then, and I cannot sell out now. I may be public enemy number one in the United

States, but I know I am teaching many people how to be free of pollution. While I struggle to survive financially, it puts a smile on my face.

They Promise Everything, I Receive a Pittance

If I had billions, I would probably spend it all doing good things for humanity. Some people are upset thinking I should have taken the money offered and then used that money to help humanity. What they do not realize is when you take the money from OPEC and promise to never do anything regarding carburetors or clean air, they mean "anything." It seems all my good inventions would interfere with their intention of keeping things off the market and away from the public, making such an offer impossible to accept. Anyway, I do like beans.

Back in 2000, I was asked to retrofit a large boiler at a galvanizing plant in Rockford, Illinois. The owners of the company were from India and had offered me a 20,000 dollar deposit to cover my costs and 40,000 dollars upon completion of the first unit on their site, plus 20 percent of their savings. I was too busy teaching to do the job myself, so I sent a technician and an assistant back to do the job for me. They photographed the job and all the components needed, and sent them back by email.

By morning, they had the designs in their hands on how to build the new reactor and new heat tubes to power the boiler. The installation took two weeks and the owners only advanced another 5000 dollar at the time. In about three weeks, I received a call saying the amount of natural gas they saved was about two thirds of what they had been using prior to the GEET.

That pretty much is a no-brainer. If their bill was 1 million dollars every three months and now it was only going to be 100,000 dollars a month, at 20 percent of their savings I would have a substantial monthly income. I was not aware at the time the same group of wealthy Indian people had conned many inventors into putting technology on their property. After four months of no payments, I phoned them. Their number was no longer in service.

After many calls, we were finally able to talk to the trustee for the court handling the bankruptcy for this plant. The court-appointed attorney informed me this company shut down, took everything of value, and returned to India. Later, I found out they had a discount with about 15 different inventions, and had no intention of honoring any agreement made in the United States.

Am I upset now? No, I believe in karma. If I had all the things which were stolen from me over the years, it would probably take 20 trucks to move it all. I know it is all baggage I do not have to move. I do hope whoever got it can enjoy it as much as I did. I have been married many times and each time, yes, I loved my wife, but when pressure was put upon them by outside sources, they all chose to run rather than fight. Being the inventor I am, I have certain beliefs and one of those beliefs is if you ever walk out my door and stab me in the back, do not come back.

One inventor I know down in Anaheim, California was telling me he had invented or reinvented a coil similar to Tesla. He first demonstrated it to very wealthy investor type people, who said they would get back to him the next day. At about 4 o'clock the next day, there was a knock at his front door; the man standing there told him to place all of his technology in a box so he could take it. Before my friend could respond, the man pointed out my friend's nine-year-old daughter, who was getting out of the back of a limousine with an ice cream in her hand. He said, "The next time you see her, she would not be alive."

With that, my friend got his equipment and put it in boxes. He cut up the coil into many pieces with carbide-tipped blades, placed all the parts in a box, and handed it to the guy at the door. He was told if he ever played with free energy again, he would lose all of his family.

I have heard these tales from over 50 people. Are they real? Absolutely. It is because of stories like this I publicly announced if anything happens to me, my family, or my loved ones, over 500 websites worldwide will be sending out plans on how to run every piece of equipment on this planet on pure water.

I am not brave, but I am stubborn. I will do what I believe is right, whether you like it or not. Persuading me with bullets, guns, or psychiatrists will not work. I use common sense. If you can prove to me with words you have a better way than I do, I may change my thinking, but until then, this is who I am.

We say time changes things, and it does. The reality of inventions is sometimes thousands of people may lose jobs over a new invention. How many thousands of new jobs could be created by the invention? This must be weighed out, and if in fact there are more jobs created than are lost, it is a good invention. However, to simply have robots to replace man with no place for that man to work is atrocious.

Imagine The Possibilities

Try to imagine for a minute there is a perfect balance on this planet. Now for me to create 1,000 BTUs of usable energy, I must take 1,000 BTUs from something, leaving it cooler. If I am pulling, I must get those 1,000 BTUs from something warmer. All things being equal, they will always balance each other, sometimes with minor inertia.

Once something begins to move, it will require less and less energy to propel it quicker and quicker. Now, consider a thrust unit propelling a spaceship or flying saucer. Once activated, it will continue to put out the same amount of force, but the speed will increase over time. It is much like what we express in the field of gravity, when something is falling and it will increase its speed. Gravity as a field is very much misunderstood, since we do not look at it as an electrical field, which it is.

We found by speeding up GEET rotations, we created an electro-gravitational field causing the engine assembly to weigh less and less as it runs faster and faster. Although at first it was just a simple observation, it became quite apparent this phenomena has a bearing on many other fields of research. It could allow us to go into outer space and return at nice safe speeds by just adjusting the amount of gravitational field around the unit.

We are probably light years away from being able to go into space at super-high speeds to perhaps overcome disastrous situations such as wormholes. Are we really so far away? No. The answers we seek are right before our eyes, yet very difficult to see. Nature supplies us with the answers to most, if not all, of our questions. We simply have to learn how to study nature better.

Most people have seen photos of a piece of straw going through a steel beam. How fast was the straw going to go through a steel beam? The answer is simple. That piece of straw had to be moving at the same atomic speed as molecules vibrating in the piece of steel. The second the straw slowed down, it became one with the steel. This is taught in the book, "Elements of Numerical Analysis."

Now here is a simple one for all of you regarding heat. In the late 1800s and early 1900s, many barns burned down because the hay stored in them spontaneously combusted and burned the buildings down. It is common knowledge you have to let the hay dry out in the field before you bail it. Wet or too green of hay, stacked or compressed, will put out a lot of heat — enough for self-combustion.

Back in the late 1960s, an old-timer took me to his cabin in Big Bear and asked me to load his truck with firewood. There was about a foot of snow on the ground around the cabin, on the cabin, and on the firewood. There was no obvious heating system or electricity in the building. Once the truck was loaded, I went into his building. It was about a 900-square-foot cabin. It had a small toilet, with a curtain around it in the corner, windows, and an entry door. This gentleman had the windows open all the way, and the door was open when I walked in.

I took off my beanie, gloves, and coat and set them on a folding chair. It was still rather hot, so I took my shirt off, leaving me with just a T-shirt. I looked around the room and could see no sign of a heater, woodstove, or any other method that could cause this cabin to be so warm in the middle of a snowstorm. I asked him how he was the heating the place. He had told me he was going to give me 50 dollars for loading his truck, and he asked me if it was worth that amount to learn how to heat a house for less than 50 dollars a year.

I was only in my teens and 50 dollars seemed like a lot of money, but I did know it was a small amount of money compared to what it would cost to heat a home per year, so I said yes. He explained he had layered up about half a bale of wet alfalfa, then a coil of poly pipe, and the other half of the bale of wet hay on top of the pipe. He said this combination was in a plastic bag sealed airtight, and was digesting the alfalfa into a compost. Coming from the center of the compost bag was hot water going to the poly pipe at about 185°. The hot water went through insulated pipes set under the cabin into a 600-gallon storage tank with a return line on the bottom of the tank back to the compost unit. This simple system allowed the 185° water to radiate under the floor to keep the building warm all winter long.

He said the only thing that operated it, since it was only for heating, was a small pump that operated on a 12 V battery, which was hooked up to a 12" x 12" solar panel. He stressed the alfalfa had to be fresh and sealed in a bag within minutes of being cut. We have tested this application in several locations of Alaska and Canada, and it works as good cheap heat.

Now this could be considered competition for GEET, but it really is not. Although GEET is fairly inexpensive to operate, it makes more sense to use the organic materials for heating if the byproduct is a good natural based fertilizer which can be added back into your ground.

Because the water is so hot coming out of the composter, there is a very good chance we will be able to remove some electrical current from a temperature differential of hot and cold water in opposite directions. The volume of water and energy produced are in direct proportion to volumes and vacuum.

Digesters can be used for many other purposes and many other sizes to dispose of waste products such as farm waste. Instead of burning it all, it could be chopped into small particles and kept moist until it can be bagged and placed in a vacuum, where it will magnify the intensity of the digesting without combusting. This will also be very beneficial to farmers, as it would reduce the amount of fertilizers used and improve the quality of their crops.

There are several ways to make a GEET. Every model of every piece of equipment on the planet, which uses fuel, will be slightly different. Then, every fuel used in those applications will have different requirements. When heating with tires, coal, or wood, you need an airtight woodstove or chamber that will help to swirl air through the vapor, delivering a good quality vapor with just the right amount of air at the right vacuum to maintain a perfect combustion.

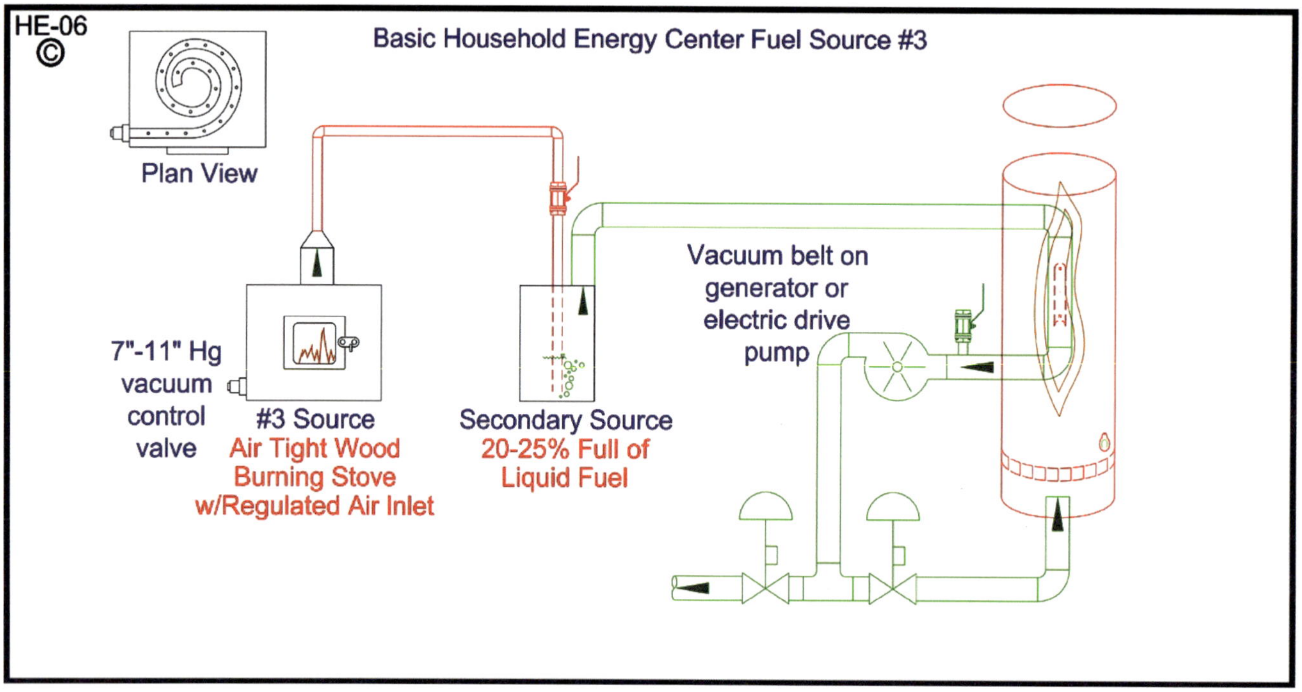

There are so many copycats, and some are very convincing, it has become almost impossible to shut them all down. If you look at all the YouTube videos, you will see many of the imposters used to work for GEET, and now claim they have a better mousetrap. Now, I am very grateful I held my cards so close to my chest and did not release the discoveries to all of mankind until I was ready. Now as I release this information, along with dates of discoveries, many people will be exposed as the liars they really are. Sometimes karma works faster with a little help.

For those of you who have never heard of GEET, this will be an interesting journey. As long as you have a clear mind and an open one, you will be able to visualize each and every one of these discoveries as they are being described. If I have failed in my descriptions and you have a better way of wording them, please make contact.

I have done my best to word things in such a way that even someone in fifth or sixth grade will be able to understand what I am saying. Back in 1994, I was trying to come up with the best scientific terms for everything. If you were a physicist, you could follow what I was saying, but no one else could understand. I try to teach at a sixth-grade level, and even some of the top scientists in the world who attended my classes have pointed out the simplicity of GEET is unbelievable.

Can you imagine being able to teach a new type of automotive class, where 12-year-olds can build a working unit after the first day? I can! I have taught children how to modify go-karts in less than two hours. Boys being boys, it has been a problem a couple of times when they decided to take a pee in their gas tank. Things like this should only be done in private. A Colorado state trooper attended one of my conferences in Denver and raised his hand to speak. I asked what his question was, and he stood up and turned toward the crowd. He said if anyone was caught urinating in their gas tank, they would probably be given a ticket for indecent public exposure. He also warned the men if any of them got stuck, he wasn't going to help them.

The shape of the rod is extremely important. When blunt rods are used, they do not allow for any quick deceleration, nor do they allow a proper magnetic centering of the rod when running. If the rod is pushed beyond the center point, it will only reduce the fuel emission by 90 percent. When putting a bullet shape on the intake of the rod area, it allows for a more even flow around the rod surface where the fuel is accelerated.

Using a flat end at the opposite tip delays the return to the proper location in the magnetic field by as much as five seconds. By grinding a cup shape in the far end of the rod where the plasma is made, it causes back pressure on the rod, forcing it back into the center of the magneto/magnetic field instantaneously.

When running a car or a large engine with GEET, for every three feet of pipe between the reactor and the intake of the engine, there is a one-second delay in engine response time. For generators, this may not be a big deal, but for automobiles and trucks pulling heavy loads, a one-second delay is unacceptable.

Many people want it all. They want the high performance, enough to rip the tires off when you hit the gas pedal. Some people are happier with five times the mileage and do not mind going a little bit slower. Most customers in the United States want more performance and more power. Our European neighbors, who pay much more for a gallon of gas, would really prefer the higher mileage. When I was growing up, it seemed to be very important to almost everyone I knew that speedometers would go to 120 to 150 mph. Since I never really needed to drive much more than 75 to 85 miles an hour, this did not seem logical or practical. Why on earth would you build a car or truck to do 120 to 150 miles an hour unless you're going to a racetrack?

If cars were designed a little bit more intelligently, so people could take off at a slower start, cruise at a comfortable speed, and achieve higher mileage, most people would be happy. You are always going to have a small group who wants to drive fast.

I remember buying a pair of brand-new Camaros back in 1986. Of course, I got the five speed, with all the bells and whistles, and bought my girlfriend a little four-cylinder Camaro with a T-top. Hers got better mileage, and I only got out of fourth gear one time. I remember bragging I had the high-speed engine, and it would do 160, or so the manufacturers said. I have to admit I never did get it above 85.

If you have the same car, available with a top speed of 90, I would still have bought it and probably saved a lot of money on gas. A certain amount of the brainwashing men get in this country make you tend to want to have all the unnecessary power.

During one of our tests, we modified an engine on a mixture of liquids getting us over 10 times the amount of run time on a generator. All the guys in my shop were upset it didn't have full power. I looked at them and asked if I could still get eight to ten times my mileage or run time, and did I really need a

5 kW generator or could I be happy with just a 4 kW generator hooked to that engine. Their common answer was it needed to put out 5 kW. To me a person with half a brain would understand that getting ten times your mileage, even if you only get 80 percent performance, is still much better than having full power, and dumping your wallet into your gas tank.

They Stick Like Glue

State of New Jersey

DEPARTMENT OF TRANSPORTATION
1035 Parkway Avenue
CN 600
Trenton, New Jersey 08625-0600

CHRISTINE TODD WHITMAN
Governor

JOHI

Date: 02/21/97

To: Paul Pantone
GEET International

From: Michael Strizki
Office Of Technology

Thank you for demonstrating your GEET technology to the Our office . Our evaluation of y process shows much promise in reducing fossil fuel emissions. Currently it appears to us from your demonstration that your claims of burning dirty fuel with low emissions is true. Please le know when your technology is to the point that we can build a prototype for testing. You still need to develop automatic metering controls to make this technology transparent to the users. look forward to joint ventures with you in the future. Enclosed is a video tape of your demonstration if you would like a copy of the data collected during the demonstration you ca contact Lyle Rawlings at 609-466-4495 . If you have any questions please contact me at 609- 4975.

New Jersey Is An Equal Opportunity Employer • Printed on Recycled and Recyclable Paper

During a demonstration in New Jersey, more than 50 top people from military, corporations, and universities from Washington D.C., New York, and many other areas were all invited. The New Jersey Department of Transportation had originally planned to have TV coverage, but camera crews were sent away 30 minutes before the demonstrations began. Our demonstration went well, and many people were thrilled by what they saw. The gentleman representing the Department of Transportation for New Jersey asked me to go outside for a private conversation.

Once outside, he explained "the facts of life," and how he and his cousins would make sure this got out to the government agencies as needed. He explained how I would need to shut my mouth, quit giving demonstrations, and he and his relatives were taking over my company. They would see to it I got a reasonable royalty. I said "No thank you", and as I walked off, he said, "You will never market your technology in this country. I'll see to it."

It appears just as firemen and other rescue people stick together as a family, and cops stick together as a family, so do politicians. Please do not get me wrong. I do not believe all politicians are crooks, just most of them. A few have called me and said they would like to see GEET used, but alone, they cannot stand against the oil companies and their colleagues supported by them.

It is hard to believe for the last 34 years I have had a solution to reduce, if not eliminate, pollution, yet I see in the news almost every week pictures around the world where people cannot see across their own city due to the pollution. When will these people wake up and realize we need to make the advancements now for our grandchildren to have clean, fresh air?

I believe in America and the people living here. I do believe they would like clean air. To get this knowledge out will require many hundreds of people teaching, not just one. I have kept a lid on how and why GEET works while I prepared copyright information which can be disseminated among the people, who may not use it for any commercial application. I want to ensure that GE, Westinghouse, and other big companies cannot keep this technology and benefits from the people.

We have for years believed this great government is trying to promote better mileage, clean air, and a good healthy environment. If this really were the case, there would not be a huge list of cars on the Internet that cannot be sold in America. Now keep in mind that Mr. Tesla, back in about 1925, had an electric Pierce-Arrow he could drive around seven days a week and it never needed to be charged. Is that really possible? Yes, it is. Is it legal today? No.

So far, I have been politically correct with what I have been saying. Some of it may seem very harsh. People should be allowed to improve their lifestyle without asking permission, unless it is obvious they are endangering someone else's life, liberty, or pursuit of happiness. If you ask a politician, his pursuit of happiness would be much greatly improved if all you got were five miles to the gallon instead of 20. We need to remove the oil companies from this equation of science and health, as well as politics.

If any oil company really cared about our environment, they would embrace new technology by investing in it. If an oil company were smart, they would invest in huge generators and run them cleanly on crude oil. Then they could sell the power they are making to the utility company. They would make more money for their crude oil than they ever did refining it for the diesel they sold to the utility company to begin with. This could all be done without the cost or pollution of refining the crude oil.

It was about 1995 when I went to San Francisco to meet with George Bechtel of the Bechtel Corporation. The receptionist told me I would not be welcomed the way I was dressed. She further explained I needed to be properly dressed to meet Mr. Bechtel. She suggested a three-piece suit. I looked down at my Levi's, T-shirt, and tennis shoes and laughed. I asked her to let Mr. Bechtel know I had come by to show we could use crude oil right out of the ground, or even water if we wanted, as a fuel. Someday, he or his company might want to meet me. I told her my dress code was much different. If he did not look casual, he would not be welcome at my home or office either, and I turned and walked out.

Over the last 37 years, many people from utility companies have asked me if I would give them a guarantee and bond for delivery of power. Since I was not asking anything for my technology other than to help humanity, I felt highly insulted. I told all of these different insurance companies and utility companies when I was finally ready to start delivering power full time, I would not need to go through them. I would just go directly to their customers and get them all the power they want for half of whatever the utility companies charged. I was not doing it for the money. I was trying to clean up the planet and maybe then they would understand you do not look a gift horse in the mouth.

Take It To A Third World Country

Let's level the playing field. The knowledge offered in North America is for the entire world, based on the income of the country where the energy is produced. Most of the profit needs to be spent or reinvested within the country where the income is made, not sent to foreign countries. Therefore, every country wishing to embrace this new technology and many others still in the wings will advance in leaps and bounds beyond those who wish to control us all. The number of inventors coming to me are in the hundreds. Some of their products are great in their own field and complement GEET. Some inventions could be used instead of GEET. Each invention does have specific applications where it would perform best.

Although I will admit it would be better for me to market in Third World countries, the sad part is most of those people cannot afford to even build the GEET units. For these poor souls, big countries like China, Russia, and the U.S. try to keep control over as much of the world as they can. They do not want this new technology; they want to utilize the old technology. It is all about control.

Sometimes looking at things from a different perspective can help you see things more clearly. If you are driving down a twisty mountain road and you know you have to go to the end of the road and turn right or left, you really cannot visualize it well until you get to the intersection. However, if you are in a helicopter looking down at it, you might be able to see the intersection long before the car gets to it. We need to mentally and spiritually get on board the helicopter and look at all the different options and paths available for the betterment of the world.

Follow The Instructions

Since the purpose of GEET is to expose the coldest possible temperature to the hottest temperature in the shortest period of time to achieve maximum efficiency, it is recommended to cool the vapor by a natural vortex prior to entering the reaction chamber. The ideal temperature for the vapor going into the reactor would be at or below freezing of 32°F. The vapor coming out of the reactor also needs to be cooled depending on the ambient condition, so the combined GEET fuel or GEET gas and ambient air temperature should be between 72° F and 80°F. This is only an approximation, and each engine will have different requirements (soon to be published).

Many people, including myself, have put GEET on two-cycle engines and been very successful in reducing pollution and increasing efficiency. Without having proper laboratory equipment, we have been reluctant to do much work on two-cycle engines, as it may still be necessary to add some lubricant to the top end of the engine after the reactor. However, since GEET burns so cool, we have not had a problem with any two-cycle engine overheating. I had a Geely scooter stolen while in Utah, getting two times the regular mileage and 95 percent less pollution.

Now, for those of you who are perfectionists, please keep in mind when you're building a prototype you don't try to get it perfect the first time. It may take several attempts to get it soundly operating for your specific needs. I have seen many people who have changed the entire GEET plans to make it look pretty. Pretty does not always work. Before you try to make it pretty, get it to work. When the plans call for half-inch copper or stainless steel tubing, it is because this is what was tested and proven.

Although it might work with one inch made of stainless steel, all with 90° bends, it may not give you the same results as simply following the instructions. Many years ago, several brands competed to be the main supplier of tissues, but overall Kleenex won. The general public gets used to a brand name, and this is the one they buy. It really is a mistake, as many companies, including top brands, will cheapen their grade as time goes on.

If the oil companies would still like to sell us gasoline, they could make a better quality, good only for GEET powered cars, and special tanks in the ground of the gas stations along with special nozzles fitted only for GEET equipment. This way they do not lose any of their money. Rather than make the attempt, they would much rather keep buying off inventors or eliminating them to keep the status quo.

Our Learning Curve Wasn't As Easy As You Think

In 2009, I had a class of about 30 people in Duncan, Oklahoma. I offered a half-price discount to anyone who became a dealer on that day, and almost everyone in the class wanted to become one. Before the class was over on the fifth day, one of the students arose and said he did not want to be a dealer under a dictatorship. He asked if he could form the GEET Dealers Association. Of course, I said yes. I would love to see the dealers working together as a team. The group elected him the president of the new GEET Dealers Association.

It was a unanimous vote that I be removed from my obligations of supplying any finished product to any dealer. I really like the idea, and welcomed their ideas to go into production of GEET products. I still insisted they all do things politically correct so no dealer would ever be in danger of being shut down by the government here, or in any other country. Well, it has been five years. One dealer demonstrated a motorcycle which was a small bike that had been motorized, another perfected a beautiful looking generator, and a third dealer came out with a kit for small engines.

The dealer with the bike wanted to sell it for around a thousand dollars to the general public, and he wanted to wholesale the bike to other dealers for 950 dollars, so they all made a profit. Hard to believe they did not sell one. The same dealer was selling custom-built product, collecting the funds, and not delivering anything. He quit.

The dealer who perfected the generator appeared at one of the conferences in New Mexico. He collected checks for many sales that also were never delivered. I still get calls once in a while from people who gave him checks, wanting to know if he will ever come out with a finished product. Since he has never even called me back in four years, I would have to say I doubt it.

We must not forget the kit. It was pretty, made of stainless steel, and the dealer only wanted 795 dollars for a kit you could mount to your lawnmower. Wholesale to other dealers would only be 750 dollars so they could make some profit. It would have only cost another hundred or two to have it all welded together once you bought this kit, and none of them sold. You could build the same thing with parts from a hardware store for about 200 dollars, and still use the parts again on another project.

The dealers thought it would be so easy to manufacture and distribute product as I had done. They were all going to become millionaires. Most of them did not realize the amount of work required.

Every time we came up with a new generator or engine, we would spend months testing and designing it so it was user-friendly and politically correct. Sometimes it would take three months to detail the generator, even though it only took us two or three hours to build it in the first place. Because I felt my reputation was important, every engine and generator was tested before it was shipped. We wanted to make sure each product was something we could stand behind with pride. Because of this testing on every engine, I could guarantee customer satisfaction.

One customer who phoned told me he was taping the call. He asked me if I could guarantee 5,000 kW of power from the generator. I said absolutely yes. He bought the generator. About three weeks later, he called me up and said he wanted a 100 percent refund because his generator only put out 3,800 kW of power. I got his address and driving instructions and drove over the next day to investigate his problem.

Once I arrived, I began testing. I found he was correct. We were only getting about 3,800 kW of power. We pulled out the analyzer and did further testing of the emissions. What we found was an abundance of oxygen coming out of the tailpipe and not enough oxygen going into the intake. To fix the problem, we simply rerouted part of the exhaust back into the intake of the engine.

I knocked on the man's door and told him that we were now getting 5,000 kW of power. He became very irate and pulled the book down from his bookshelf right there by the front door. He opened it up and pointed to a chart. At 11,500 feet in elevation, the engine was not able to get enough oxygen to put out 5,000 kW. His electrical engineering book clearly showed according to science, the generator should only put out 3,750 kW. It was obvious he was trying to get a free generator, but it did not work.

How To Lose A Bet

Time after time people have tried to take advantage by making bets they are sure they cannot lose, and they have lost. One of my favorite occasions was a man who owned an auto parts store in Auburn, California. He had seen me buy so many strange parts over the years and his curiosity got to him. He asked me, "What on earth could you be building that you need parts for a VW, Chevy, Ford, and all the other strange things you've been buying? They could not possibly work together."

"An engine that runs on almost anything," was my answer.

He pulled his keys off his hip and threw them on the counter in front of his three adult sons. Then at the top of his voice in a crowded store, he said, "If you show me an engine that would run on almost anything, you can have this store."

I asked him to follow me out to the parking lot, and I invited all the customers who were listening to join us. In the center of the parking lot on the ground was my engine. I looked around and I saw people coming from the third shop in this strip mall. Some were coming from a four-wheel drive repair shop. I asked one gentleman named Doug what was the worst liquid he had in his repair shop, and could I please have a cup or two of it. I handed him an old coffee cup.

He returned a few minutes later and I poured this great stuff into the fuel canister as he announced it was a mixture of oil, tranny fluid, and some rainwater. The leaves were for texture. Everyone laughed when I poured the fuel into the fuel tank, except me. I just smiled. I mentioned how Bob had said I could have his auto parts store if this engine would run on what I just put in the fuel tank, and Bob nodded his head yes. I asked if anyone else wanted to bet me. Three people offered 20-dollar bets it would not run.

With this, I pulled the rope and stood back about 10 feet, laughing because the engine not only ran but it also ran smoothly. I went over to the three guys who bet 20 dollars each and took their money and then turned towards Bob. As I looked around, I saw the door to the auto parts store closing. Bob had run off. His three sons were saying he was only joking. I ran into the front of the automotive parts store, just as Bob was running out the back door. When I looked at the counter, the keys were not there.

His sons came back into the store, with humble looks on their faces, realizing their father had really blown it with his big mouth. They asked me what my intentions were, and I said I just wanted to have some fun with their dad and I had made 60 bucks on the side. The boys decided the 50 or 60 dollars for the parts I had on the counter were free, if I was not going to take the store for the bet. I told the kids I did not intend to take their dad's store, but it was sure fun and I wanted to rattle Bob for a few more days about wanting my keys, and we all laughed.

The next three days Bob stayed home and did not answer the door when I knocked. I went back and told his sons I believed I had had enough fun. They thanked me. The shop had been running very smoothly without any pressure since their father hadn't been there. They also said this might have taught him a lesson on keeping his big mouth shut when he did not agree with things people said.

I attended a conference up in Blaine, Washington years ago where there was a scientist with several PhDs who had been brought in to play devil's advocate. No matter what tests you did, they would tear it apart and say, "You did the test improperly, therefore, it was invalid." I listened to three days of this, and many of the scientists who were making their presentations left the podium in tears or on the verge of tears. I did not want this to happen to me, so when it was my turn to take the podium, I walked up with the reactor rod and a compass and a felt marker in my pocket.

The one professor who had attacked everybody was sitting up front to my far left. I approached him very slowly, smiling, and I asked him if he would mind being my helper during this demonstration since he looked like such a smart person. He did not care for my comment about him looking smart and he grimaced but said yes. At this point, I looked at the rest of the group and asked them all to pretend they were my students for my demonstration. To begin with, I asked all of my students to repeat after me.

I smiled and said, "North is up." I motioned with my hand and they all spoke up at the same time saying, "North is up." I then turned towards the professor, held out the rod in my left hand, and asked him where north was. He said he couldn't tell without a compass. I raised my hand for the class again and they all said, "North is up."

I could see by the expression on the professor's face that he was not being entertained. He was very upset. I pulled out a compass from my pocket and held it up next to the rod. He then said, "North is at the top of the rod.

I slid the compass down the rod and showed him that south was at the bottom of the rod. Then I put the compass away in my pocket. I handed him the felt pen and asked him to put an X where he found north. I turned the rod upside down and asked him where north was. Of course, he pointed to the bottom of the rod where the X was. I stepped back, shook my head, and raised my hand again for the class to say, "North is up." Again, I pulled the compass out of my pocket and held it up to the rod. The compass indicated north was up.

The professor look so confused and rattled it was hard to not laugh at his index finger pointing at the top and bottom of the rod back and forth. I went back to the podium. I said, "Never mind. Obviously the professor can't help me; he doesn't even know which end is up." With this, all 50 people in that room laughed, further embarrassing our professor from Stanford.

If you learn nothing else from the last section of this book, know sometimes it is best to keep your mouth shut or you may insert both feet. The things you find in science books are things clearly understood with some theory behind it. Science is a weird field. Many scientists feel a theory is more important than a physical working model. Inventors are not scientists. They would rather have a working model than a theory any day of the week.

I Have Always Been This Way

It was not until I was about 22 years old that I really got fed up with people marketing my ideas and claiming they invented them. I tried to find a way of stopping them from stealing ideas in many ways. I came up with a couple of really good ones that worked. If it was a piece of clothing I had designed and someone would come up and say, "Where did you get that neat looking shirt (or pair of pants)?" I would always say it was from such and such a store on Rodeo Drive in Beverly Hills. My favorite was a T-shirt with hidden pockets. I always wondered what happened at that store when people came in wanting to buy T-shirts in a store where they sold 80 to 500 dollar shirts.

In the late 1970s, I invented a 25-caliber log splitter. By the 1980s, I become very efficient at using it and had fun doing it. If I was driving around in a suit and saw guys splitting firewood the old-fashioned way, I'd pull off to the side of the road and ask them if they wouldn't mind my splitting just a few logs. Usually these well-built guys would say sure. I would then proceed to pull out the gun, put in 30 cartridges, walk over to where they were, and blow a few logs apart by simply pulling the trigger. When they said they wanted to buy one or asked where I bought it, I always told them Brown's Hardware in Sacramento, California.

Sure glad I did not work at Brown's Hardware. I sent people to Brown's Hardware for instant paint and water, a siding tool that installed eight pieces of siding at a time for horizontal siding, or the sheet rock removal tool which would allow you to remove a full sheet of rock or decking almost as fast as your arm could pull it off. However, the sign over their door said, "If you want it, we will find it."

For the last 37 years, I have had people give me plans for over 50 different types of engines that would replace the regular internal combustion engine in your car. Many of them work, and I have seen the prototypes. I cannot imagine any government willing to allow other people to use these engines because of their phenomenal performance.

For many years, inventors have been pushed aside by our legal system, and today I have two such cases going, one in Oklahoma and one in New Mexico. I am not holding my breath I will win either one. In Oklahoma in 2010, the judge signed an order for me to continue building my home and school and labs, and after spending almost half a million dollars, the same judge ordered I stop all construction and pay rent on property I thought I was buying. I am sorry. No lunatic would have ordered me to stay on the property and continue building, and then ask why I continued to build!

As an inventor, I can invent anywhere and I can be comfortable anywhere. Sure, it is nice to have a house with air conditioning and everything you want for luxuries, but I believe my ancestors were much more comfortable with a small hut, growing their own foods, and hunting for food when they needed it. I cannot blame all the corruption on modern-day man, when it goes back to the beginning of time.

I have heard it said that only the strong survive. This does not mean muscle strong; it means power. And power is money, or in some people's eyes, it's gold. For me, power is knowledge.

Every day brings something new. Many inventors bring me their products for my opinion. It breaks my heart to have to say it has already been invented or patented. Some inventors do not even know how to do a patent search and do not trust attorneys — which is understandable — and proceed to raise money for their new idea.

One such inventor told me he had invented a new device which would fit into the trailer hitch on the back of most vehicles. It was a flat bed with a small ramp to hold wheelchairs, small motorcycles, or other equipment. I told him to give me a day or two to research. What I found was very upsetting for me, and I knew it would bother him a lot; it had already been patented for many uses. I found his office to warn him not to go into production, and his secretary said she did not want to bother him since he was in a meeting with some investors who were going to fund the company.

I stopped by a neighborhood store and picked up several magazines which I skimmed through. I found the patented version of what he wanted to build advertised in many magazines. I then drove to his office. When I arrived, the secretary told me I could go right in and that he was expecting me.

I asked the inventor if we could talk privately for just a moment. He said absolutely not — his investors had a right to hear what I had to say. With this, I opened up the magazines to the ads, one after another, and laid them in front of them. Most of the ads said "patented" or "patent pending." The investors, after glancing at the magazines, began demanding their money back. He opened another door going into the shop. It was stacked floor to ceiling with the units he had already made. He told the investors he would simply sell the products already made, and pay them back what he owed them.

After that, the inventor's investors were mad at me, and so was this inventor, for some reason. They all thought I was responsible for his incompetence in handling the situation. This is one of the reasons why many investors like to look at patents more than 18 to 20 years old. They know the inventor is already out of the picture and they can copy anything. Nobody can stop them and they do not have to pay any royalty.

What About Your Dreams?

How many types of dreams are there? Ask a psychiatrist or psychologist and they will argue and say three or four. They are wrong. I can validate seven types of dreams, and I have done so for a few psychiatrists. The problem has been since I do not have a PhD, they cannot use my knowledge and their teaching until the time when I might acquire such a PhD. I am very selective as to whom I teach this technology for fear of what they might do with this knowledge. However, I will highlight some of the types of dreams without going into tremendous detail.

Type I is a total imagination dream of being in a beautiful place or with certain people you have never met. It is usually quite pleasant. If you wake up from this type of dream and think about it for a few minutes, changing some of the things you'd like it to have been, you can go back to sleep and have the same dream reoccur, with it happening the way you would like it to from thoughts of your conscious mind.

In Type II dream, you are seeing things appear to be much more real. While in the dream, if you are normally wearing jewelry you never take off such as a ring or watch; you will try to make an effort to see if it is with you in this dream. Many times people will see an accident before it happens, as I have. The dream is no more than a preview of what could happen if you allow it. To stop something bad like a car accident from happening, you simply wake up after the dream and write it down. This way, it will be brought to your conscious level. If you do not, you may get to experience the drama in real life. If you choose this path, then you will be saying "déjà vu!" I prefer not to have the pain and suffering, so I wake up from my bad dreams and write them down.

Type III is the same as the type two, except this time, you know it is not you in the dream by seeing a different watch or jewelry than what you own. With my personal experience, I have found this type of dream occurs if I am locked into a person's frequency and I am sharing an experience they are about to go through. If I choose to be prepared for such tragedies, I must wake up from the dream. Normally, I have found these dreams only occur within 72 hours of being in physical contact with the person or an object person had in their physical possession and left their frequency upon. I could expound upon this with a preponderance of evidence, but it is not all directly related to GEET.

Type IV is a type of dream occurring when one is given a vision prior to an event like a tsunami, earthquake, tornado, or any other major event which would normally cause an emotional declination in one's ability to function normally. If you wake up from this type of dream and write it down, you will not be distraught in any way when the event occurs. This can be a great benefit if a tragedy occurs, and when everyone else is very upset and emotional, you are very calm. This allows you to continue to function calmly and normally, when others are chaotic and confused.

Type V is gaining knowledge through your dreams and a great way to gain an education without having to go to school. In my dreams, I must have built hundreds of magnetic motors that were all failures. When I was 6 years old, though, I woke up from one dream knowing how to build a magnetic motor that worked. At 35, I finally got around to building the one motor I knew would work.

See if you can figure out what the other two types of dreams are, as I do not address them in this book. I am handing you a great deal of knowledge, and not all of it is complete. There is a reason for this when teaching. It is to teach you to use your mind and expand from the knowledge you read to seek additional knowledge.

Back in the mid 1980s, I tried to explain to a very close friend she had the abilities I do, but she had not developed them. I asked her every day to try to transmit a thought onto a blank white wall. After many months, I walked into my office and saw in bright blue letters about six inches tall the message "I love you" with a smiley face. She had done it. It brought me great pleasure to know she had found the way of leaving that frequency transmitting on the wall until I received it. Then for the next three weeks, she left messages everywhere.

Keep in mind we all have frequencies. We are electrical energy and can transmit thought to one another without opening our mouths. If the person you are in communication with has a weaker capacity than you do, you gain the ability to be able to enter their thought process and hear what they are thinking. If you learn how to use this, do not. It can be very devastating knowing the person you are with has other intentions than the ones you believe. I know.

If You Build It, It Will Run

When I first invented GEET, I had a slight fear I was mentally creating a personal energy, causing the engine to run with no pollution. I had some concerns which were cleared up when many other people from around the world were able to build the same engine and get the same results without my ever touching the plans they were working with or even talking to them even one time. Still, a few scientists insisted somehow I was able to transmit. The engine should work, and therefore it did. My fear of this really happening was confusing, when there was one person who kept showing up in different parts of the United States for demonstrations. Every time this man showed up, nothing would run. On one occasion, the engine was running and he came around the end of the building. Just as he became close enough to see the engine run, it died and we were unable to restart the engine until he left.

It took about six years, before this engineer actually got to see the engine run when he showed up. On that day, he was with several other engineers who all wanted to see the engine run and it ran flawlessly. I cannot come to any conclusion I can explain. Then again, scientists have been trying to figure out why a bumblebee flies. According to all of their known science, it cannot.

Don't Tell Anyone The Truth

In about 2010, a writer from Germany contacted me and asked to write a book about GEET. Of course, I said yes. He requested names, addresses, and phone numbers of anyone in Europe who may have had first-hand experience with it. He also asked why this was not on all the new cars coming out. I tried to explain the oil companies do control governments, not just in the United States but also around the world in partnerships with local governments.

He expressed disbelief, but said he would still do some research before making a final decision. He then proceeded to phone a few car companies, and what he heard surprised him. Several of the companies told him they had already tested GEET. It worked far better than the law would allow. One of the companies told him that it might be 50 years before the GEET could be used on vehicles for the public. They recommended the inventor consider military contracts, as they would use the invention immediately.

He phoned me back to inform me of his findings, and he didn't sound happy about it. He has written many books, but most were based on current research. He was now able to understand why, after all these years, GEET was still not on the market. He said the only way to get it there was through military application, and would I consider it. My answer was simply no.

The writer decided not to do a book about Paul Pantone and GEET. He did wish me well in my endeavors and said I should leave the knowledge for my children or their children. He said someday in the future, car companies and governments might want clean air and better mileage. For now, it is not politically correct.

I have had television people come right out and tell me whatever they are told to say is what the public believes. Any story which might hurt the income of any other supporters or directors will never make it to the news. During some of my demos, reporters have shown up to cover the story, and they bring their own qualified, educated people to tell them whether the product really works before they air the show.

In one case, the engineer would not even wait until the engine was running before he told his friend it was impossible and he was wasting his time to even film it. When I asked him how he could make such a statement before I even started the engine, he said any idiot who knows anything about cars and engines would know no engine could possibly run on soda pop and pickle juice mixed with sugar. He went on to explain if the engine did start, it would be because of a hidden fuel tank inside the cylinder of the piston, or hidden under the floor. With these comments, I pulled the rope and start the engine.

"See I told you so," he said, jumping up and down and pointing at the exhaust. "There is only clean air coming out of the tailpipe!"

I am sure to this day he believes he was right. Just like religion, we are all entitled to our own belief.

Back in 1996, I met a doctor, an M.D. whose name was Dr. Rick. He was in his 90s and had become famous early in his life when he was pushed out of an airplane with two bags over the South Pacific. He was the first M*A*S*H unit. He told me the story behind the two bags and it was very sad. One bag was for the enlisted man so he could treat them. The other bag was for the officers so he could cure them. He told me there have always been cures for every disease, but he wasn't allowed to publicly talk about it.

In every scientific or technological field, there could be advancement if desired. If I have the right backing, we can probably put most of America back to work within the next year. Most people today collecting welfare and public assistance enjoy their monthly check. However, there are many who would much rather be productive, if there were jobs. You can make a difference. Create something new just as I did. Then get it out to the world in whatever manner you need to get people a good job and a satisfying life.

In 1995, a gentleman phoned me and wanted to bring scientists to my home to prove I was a fraud. I told him to bring his friends and we would greet them with open arms. When he arrived, he introduced me to two engineers from MIT who were there to validate or expose me. For the next three days, we satisfied all the tests they wanted. On the final day, my sister-in-law Patty asked the old man what he did for a living. He smiled and said he was a retired nuclear physicist from the great People's Republic of China.

When he said retired nuclear physicist, my ears perked up. I asked him, scientifically speaking, what he thought of GEET. He reached across the table, picked up one of the rods, and smiled. After turning the rod back and forth, he finally gave me his opinion. He said, "What you are doing for 25 cents, with parts from the hardware store, we are doing with big, seven-mile underground facilities for billions of dollars." He went on to say, "The big difference was not the price, but it was the result. Ours only works for milliseconds and yours was like the Energizer Bunny. It keeps running."

Before leaving, the physicist said I had a harder job in front of me than he had in the last 30 years. He said no scientist would ever stand up and say the GEET works. None of them would want to admit there is something so simple they could not figure it out. He also said no one with 15 to 25 years of education would be willing to say a ninth-grade dropout surpasses them in knowledge.

The Auric

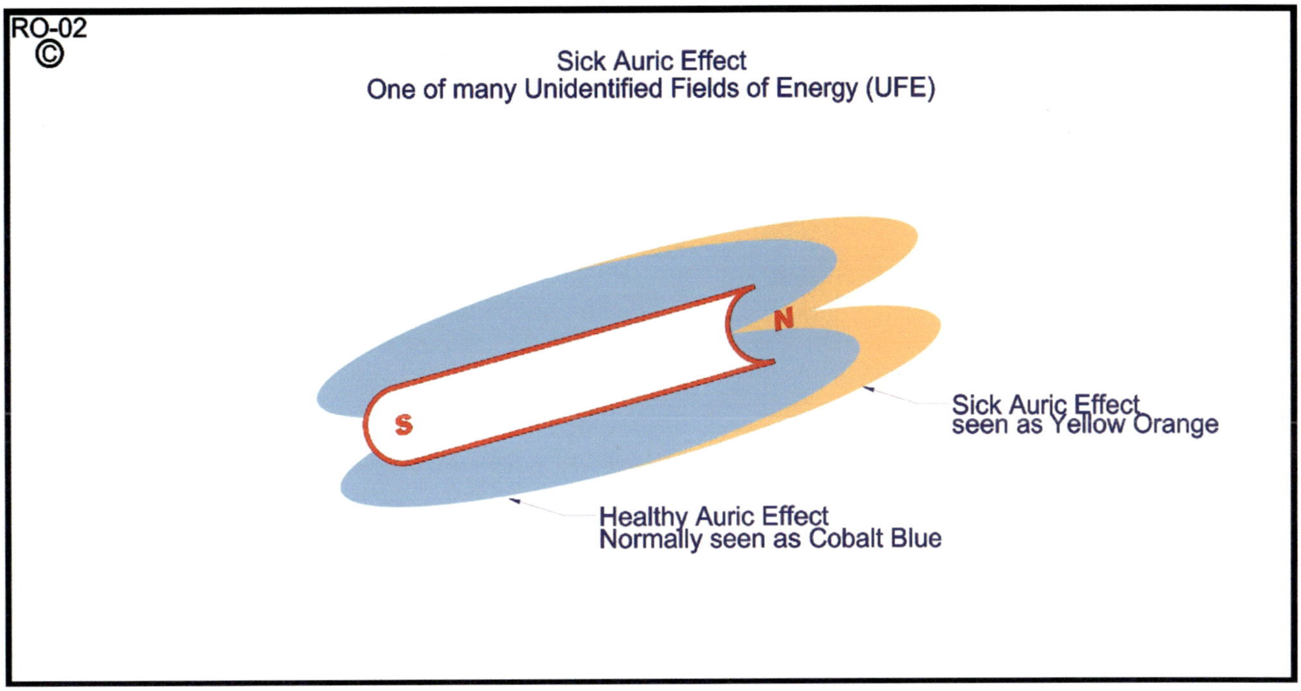

Not everyone has Kirlian photography machines, so using a compass is the next best thing to check the rods. When you hold the compass to the uppermost end of the rod and you find north is above the rod, not on the rod, this tells you how long the rod should be for the proper reaction. It shows where north should be for the heat required for the plasma field.

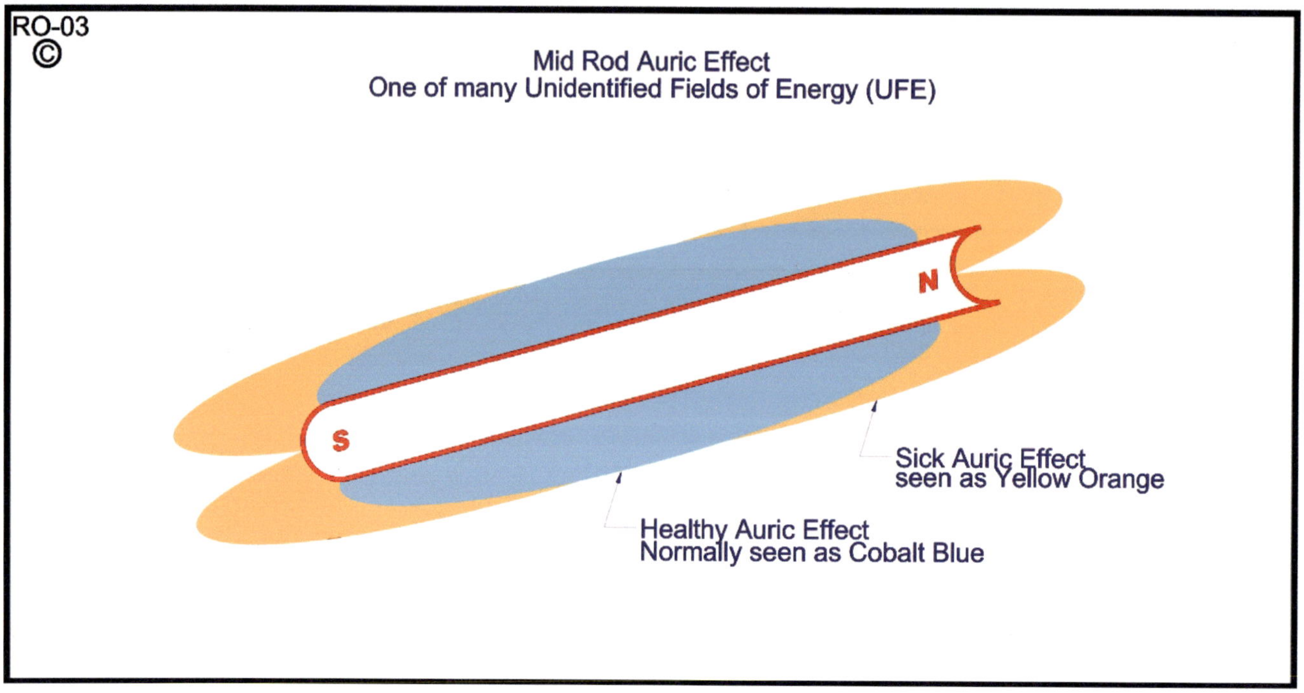

If you find north and south is beyond the end of the rod, then your rod is too short. Determine the length where the compass shows north and south are and make the rod match that length. The first time the auric field was discussed on stage, I managed to upset many scientists. I had referred to this field as an aura. Several scientists were getting up and ready to leave the conference.

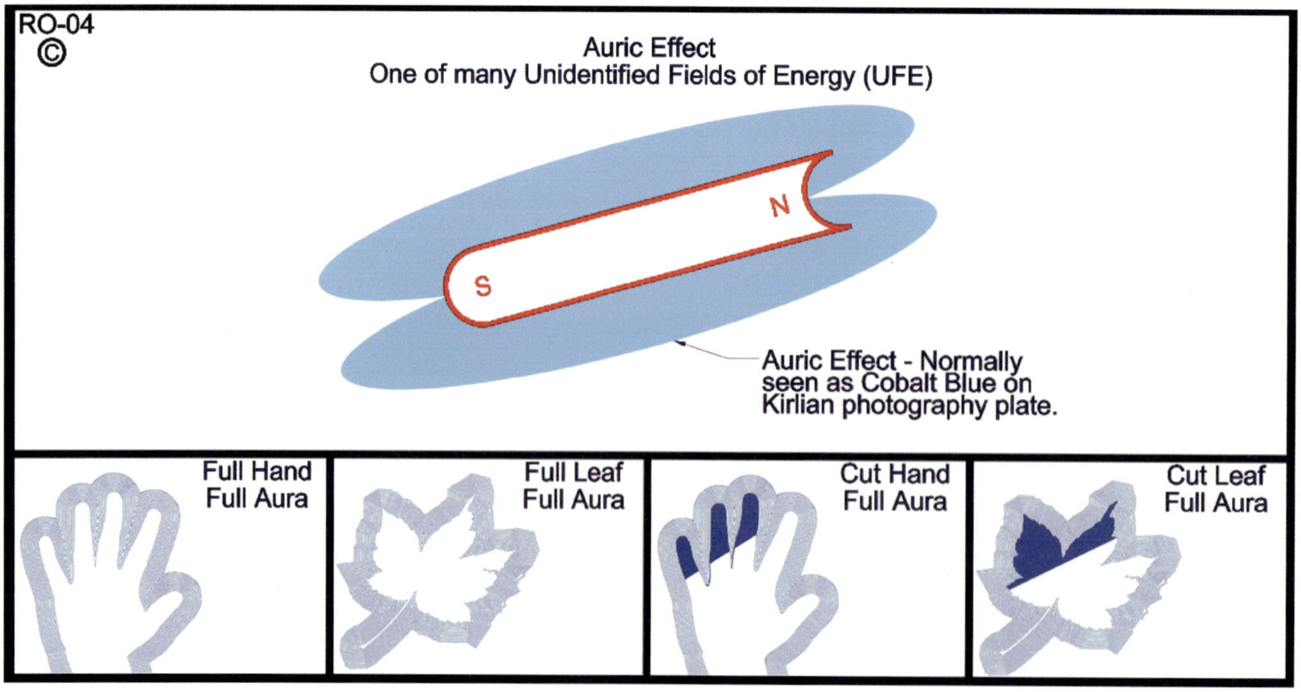

Before any of the scientists could get to the door, I stopped one of them and asked him what was so upsetting to make him want to leave my presentation.

"Only plants and animals can have an aura," he responded.

"How about if we call it an auric field?" I asked. He thought about it for a second.

"That would be all right," he said. "We wouldn't have to reprint all science books by calling it an auric field."

We Tried Many Things

When we released free plans on the Internet back in 1999, we were flooded with emails. Some days we would receive as many as 300 or 400. One of the more popular questions was what type of material could be used to make the rod. We would tell people to just use mild steel and left it at that. Rather than telling the people they could use brass, copper, stainless steel, or gold if they wanted, we just stuck with the simple answer: cold rolled steel.

Along the same lines, some of these people would ask if they could use a flat rod or triangular rod or even a square rod and just twist it. We tried every shape we could think of and it did not matter. None of them worked as good as a round rod. During tests in our shop, we even ran the process through an all glass reactor. The exhaust pipe, the reaction chamber, and the rod were just glass. We got an education doing this, but it did more harm than good with the students. Many of these overeducated people ran out of the room when they saw lightning bolts going up and down the rod. Some left screaming, "It's going to blow up."

I will say building the rod out of glass was fun, especially after it was broken in. If you really want to have fun, hand a compass and small glass rod to a scientist after it has run in the GEET reactor. The rod has a north and south end. When you flip the rod end for end it has north and south again, but always in the same direction. The real mistake we made was when we handed the rod to scientists and let them check the rod themselves. In many cases, they would drop the rod when they saw the field flip. I got tired of building glass rods and watching them all get broken. We then proceeded to tell everybody to make the rod out of mild, round, cold rolled steel.

We have tried just about everything we could find as fuel. The only two things we found that would not work in a GEET were synthetic motor oil and antifreeze. If you make this mistake, it is easier to replace the chamber and rod than it is to clean it up. These substances leave very messy black burned-on carbon deposits you cannot easily remove.

RO-07
©

Progressive contamination will eventually create a NON-REACTOR

The same type of contamination could occur if the rod is too long for the specific fuel. It will only accumulate on the excess length of the rod. The breakdown occurs on the correct length of the rod, but beyond the needed length the output GEET gas will recombine and deposit on not only on the extra rod length, but also on the inside walls of the reactor.

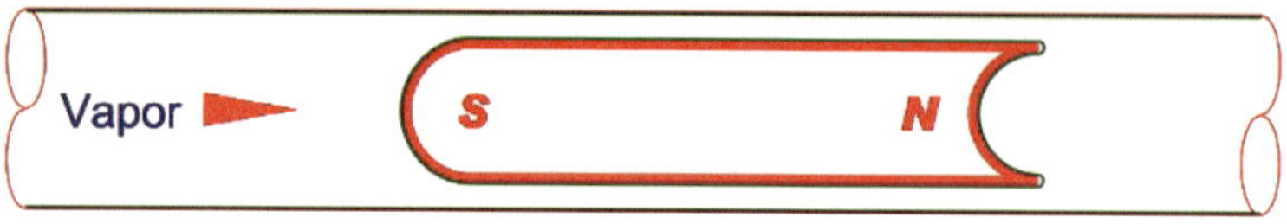

STANDARD ROD DESIGN

Back in the early 1990s, we still made most of our rods with square ends, and we also welded three tips equally spaced around each end. We tried to force the rod to be in the center of the field to disperse the air and fuel equally around the rod. We were unaware it was grounding out, canceling several of the electromagnetic gravitational fields.

RO-08
©

The University of Berlin discovered 32°F at the center of the North end of the rod and 120°F at the South

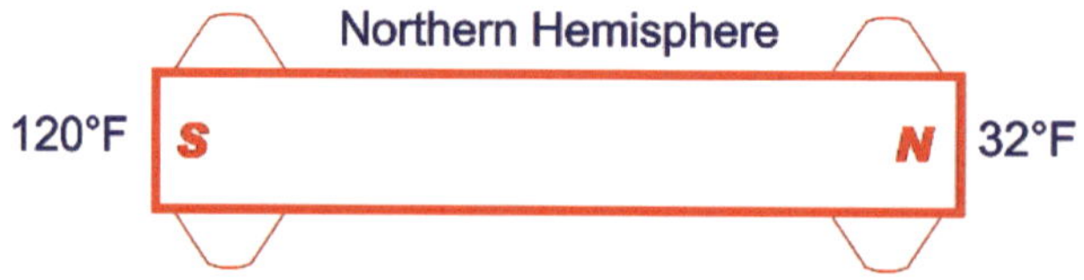

Findings at a University in Australia - we discovered exactly the opposite.

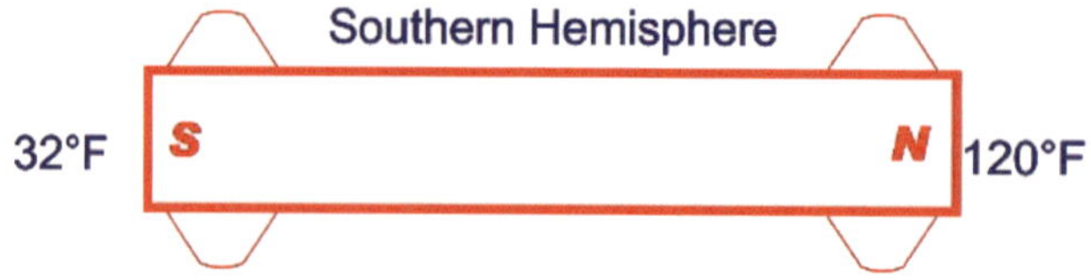

Additional proof of Double Axis and unidentified energy movement & energy being absorbed and emitted.

The free plans we posted on the Internet all showed a 12-inch rod, but we did not explain the 12-inch rod needed to be cut to size. As seen in picture RO 09, sometimes a rod will be too long and the compass will show you how long it wants to be for the fuel it just modified. All fuels have different length rods. Using a rod that is too long to do your initial run will give you a magnetic signature for the proper length you need for a good plasma burn.

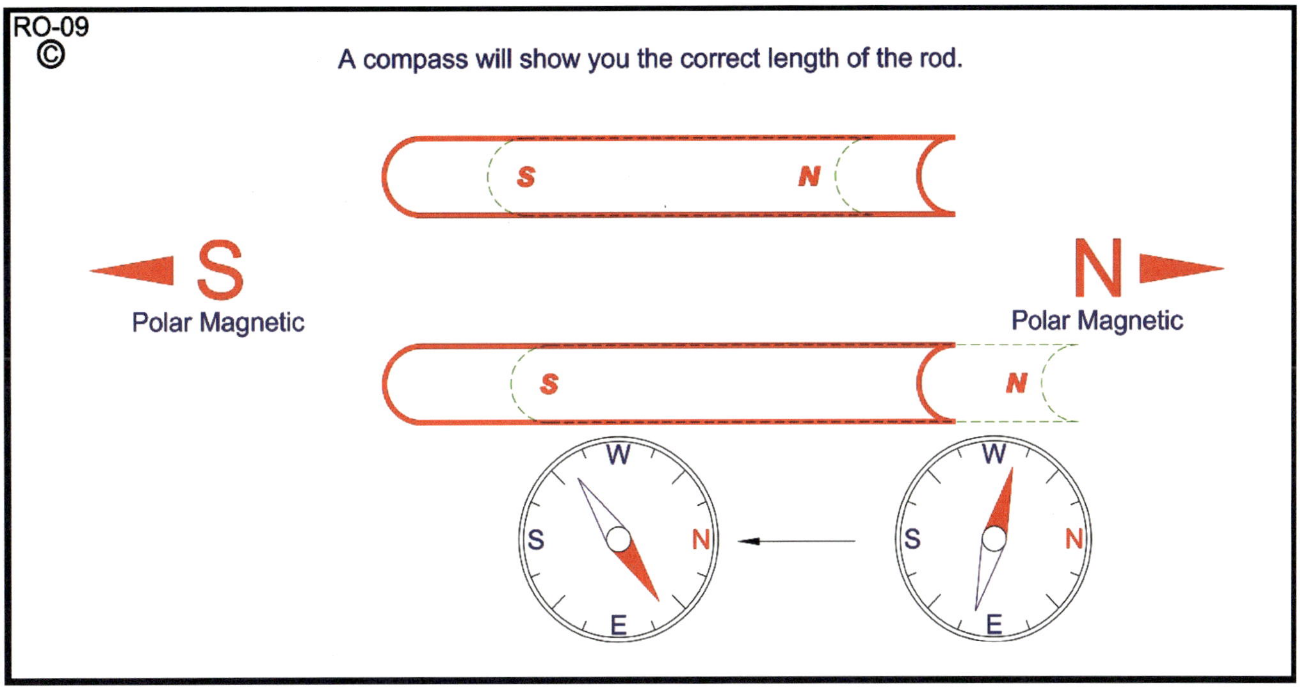

Fields Of Magnetism

In the early 1800s, a German mathematician named Carl Gauss devised a clever method for measuring the strength of a magnetic force. His early magnetometer consisted of a bar magnet suspended in midair by a string. While it did indicate the direction of the magnetic field, it offered only a crude measurement of the strength.

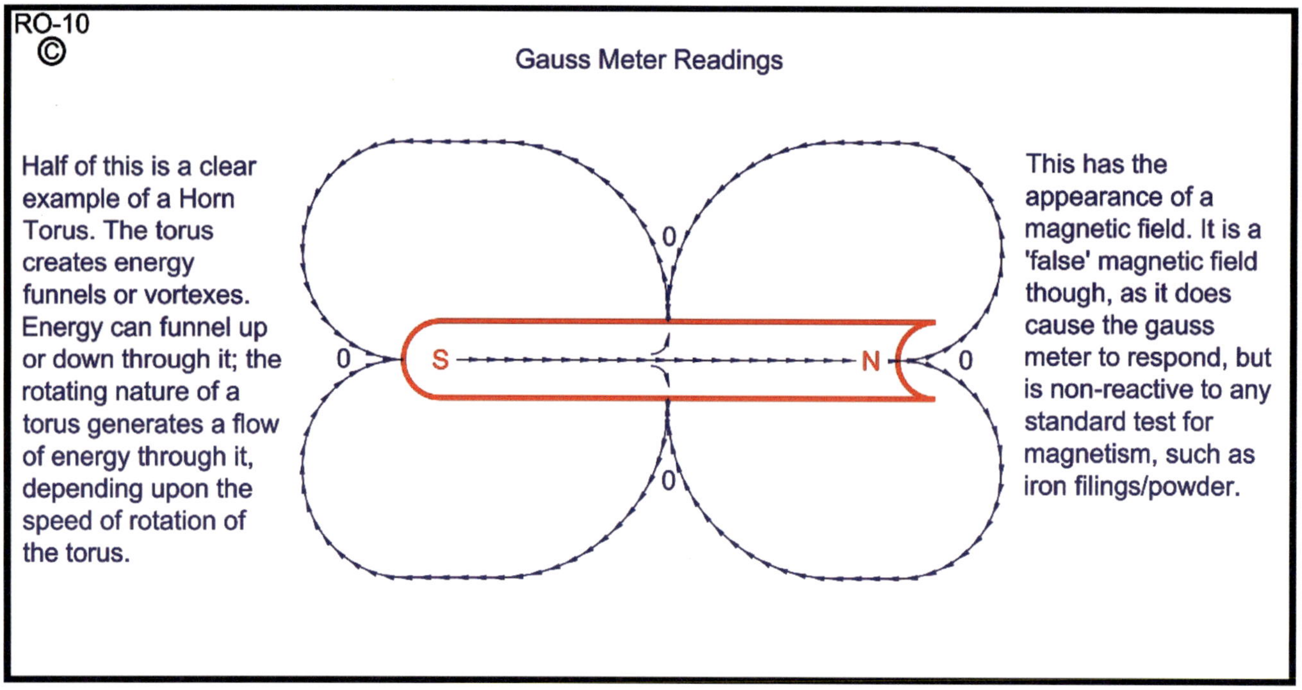

The modern Gauss meter comes in two varieties: a scanner meter which measures just magnitude at a given point and a vector meter which also indicates the direction of the field. These meters can be aligned to make a vector Gauss meter. Magnetic fields are measured in Gauss units. The Gauss unit is also used to measure the Earth's gravitational field.

A physicist who came to our school had a full-time job working for NASA and brought his own Gauss meter. Lo and behold, the Gauss field he found emitting from the GEET device replicated the Gauss field of planet Earth. Further testing showed a second Gauss field at about 10 feet out from the operating GEET unit, which gave the same basic effect as the Van Allen belts.

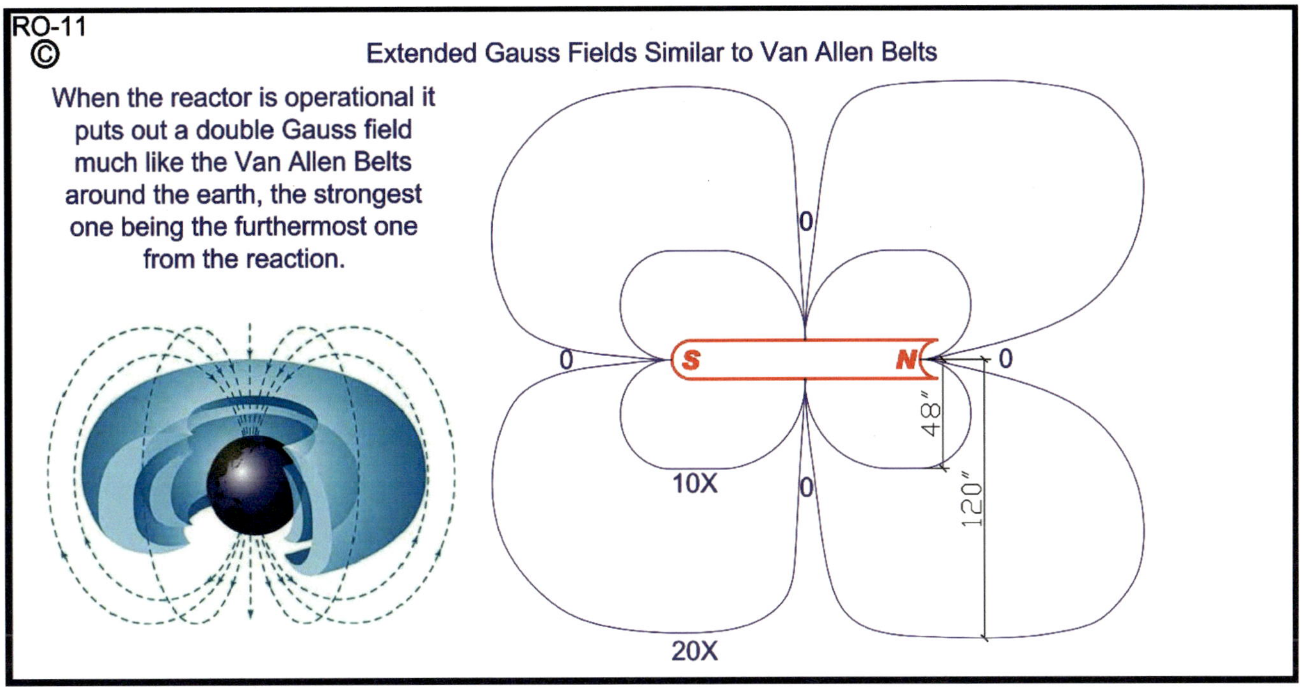

When I first told scientist we had a similar field to the Van Allen belts, several scientists rushed to correct me. They explained to me the Van Allen belts could not occur in atmospheric conditions. Boy, were they wrong. Now using GEET, universities can study the effects of the Van Allen belt and Gauss fields within the classroom without the need of orbiting satellites.

All Rods Are Not Hot Rods

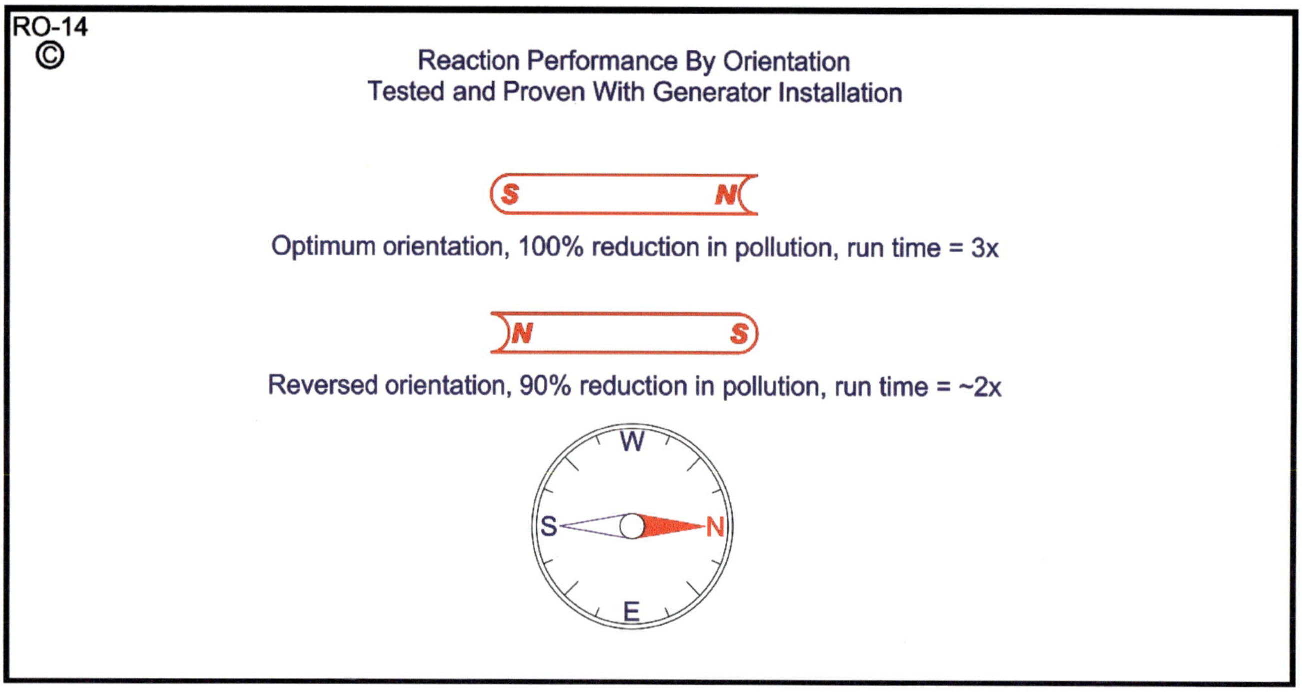

When running a rod in a horizontal position, you get maximum efficiency if the cup of the rod points toward earth magnetic north. As an example, an automobile equipped with a GEET reactor properly aligned north may get three times its normal efficiency. When you turn your car south and the rod is now backwards to the earth's magnetic flow, you may only get double the efficiency.

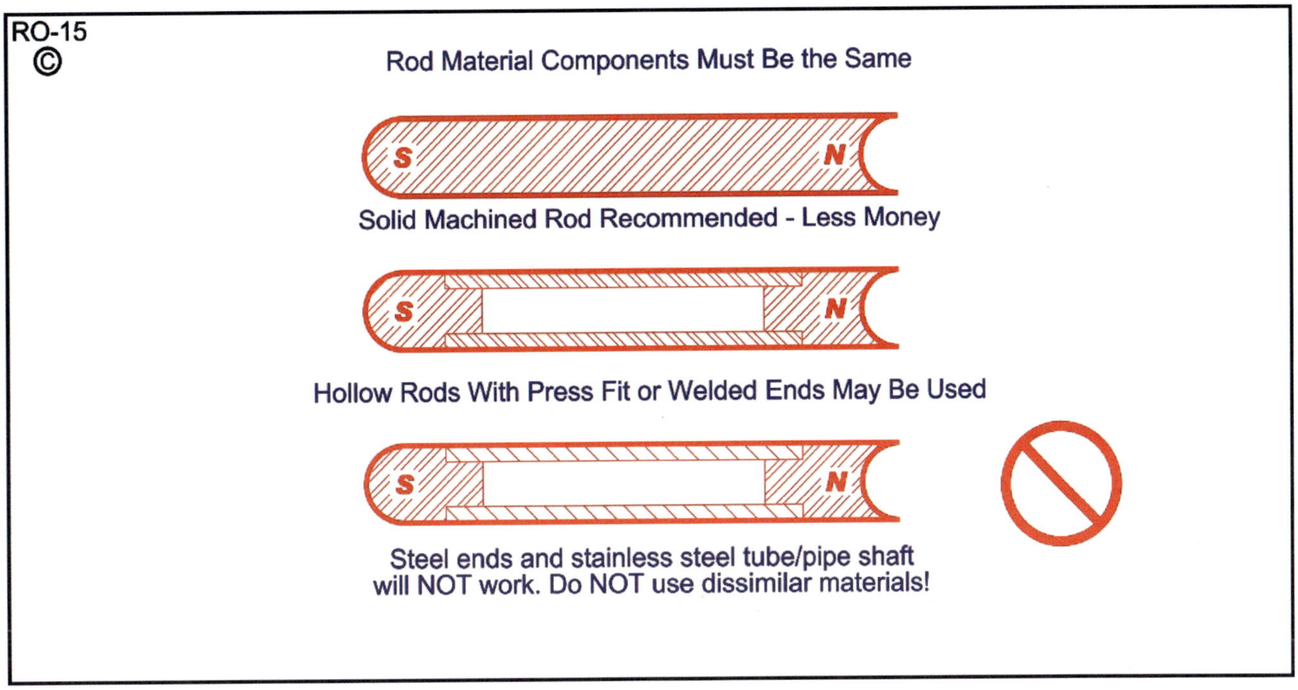

I do not know how many machinists have gotten ahold of me in the last 20 years wanting to make the rods. They have quoted prices to me in quantities ranging from two dollars a rod to 500 dollars a rod. I am not one to waste money, so I buy rod stock in three to four foot lengths at the local hardware store and cut off just what I need. Each small section of rod cost me about 50 cents or less. Of course, a machinist is going to want to machine the rod out of a bigger piece of metal and make a tidy profit. Some have suggested exotic metals. Others have suggested bi- and tri-metal reactor rods, precision cut and pressed together.

While some of our R&D team has tried gold or silver plating, and even glass, good old cold rolled steel is still the cheapest and works fine. At one point, I decided to try some titanium. I located a company back East who would cut and shape the metal to my specifications. I sent them a sketch, and the following morning I had a visit from several men in military uniform.

When these gentlemen arrived at my school in Idaho, they asked for Paul Pantone. I introduced myself and asked how I could help. They wanted to know why I was trying to order titanium bullets. I tried to explain they were not bullets — they just had a bullet shape. Then I was asked again why I wanted titanium-shaped bullets. I simply answered I needed them for my reactors. I could tell by the looks on their faces things were going from bad to worse.

Rather than sticking my foot in my mouth any further, I asked them to follow me into the shop and I would show them what I was making. I explained how the bullet-shaped rods fit into the reactor so fuel could be accelerated to make a lighter fuel which would burn better in an engine. After they saw the engine running on some soda pop, coffee, and gas, they all laughed, and one of them made the comment, "Soda pop is really expensive."

About a week later, my new titanium rods arrived and we tried them. I was not impressed, except for the bill that came with them. They were extremely expensive. We tried to calculate how much more it would cost to use titanium rather than cold rolled steel for the less than one percent gain in efficiency. It came out to be about 140 years for a payback. Based on this number, we decided titanium was really out of the question for the small improvement. We would have done better to leave our money in the bank and make a profit.

One of our research teams did build a 24-karat, gold-plated rod. This also gave us about a one percent improvement. It also was not worth the money spent. I have been trying for years to keep it simple. The simpler the materials and system, the easier it is to teach. When teaching, I try to use examples like the movement of the clouds so students can visualize what I am talking about. Sometimes I use the movement of the air through the trees to explain some of the principles. Other times I used a flushing toilet showing a vacuum pulling the water and other material from the toilet into the drain using a very visible vortex, created by the water movement down the drain.

As The World Turns

As the Earth is spinning from the west to the east, energy from the sun passing through our atmosphere strikes the earth. As it penetrates the ground or crust, the energy reverses direction to the direction of the planet rotation. Once it enters the magma that is moving in the opposite direction, it reverses spin again, creating heat from the energy. Any excess energy is sent out the North Pole. Some of these energy particles reentering the atmosphere will cause the aurora lights. This energy cycle is the same in the GEET reactor.

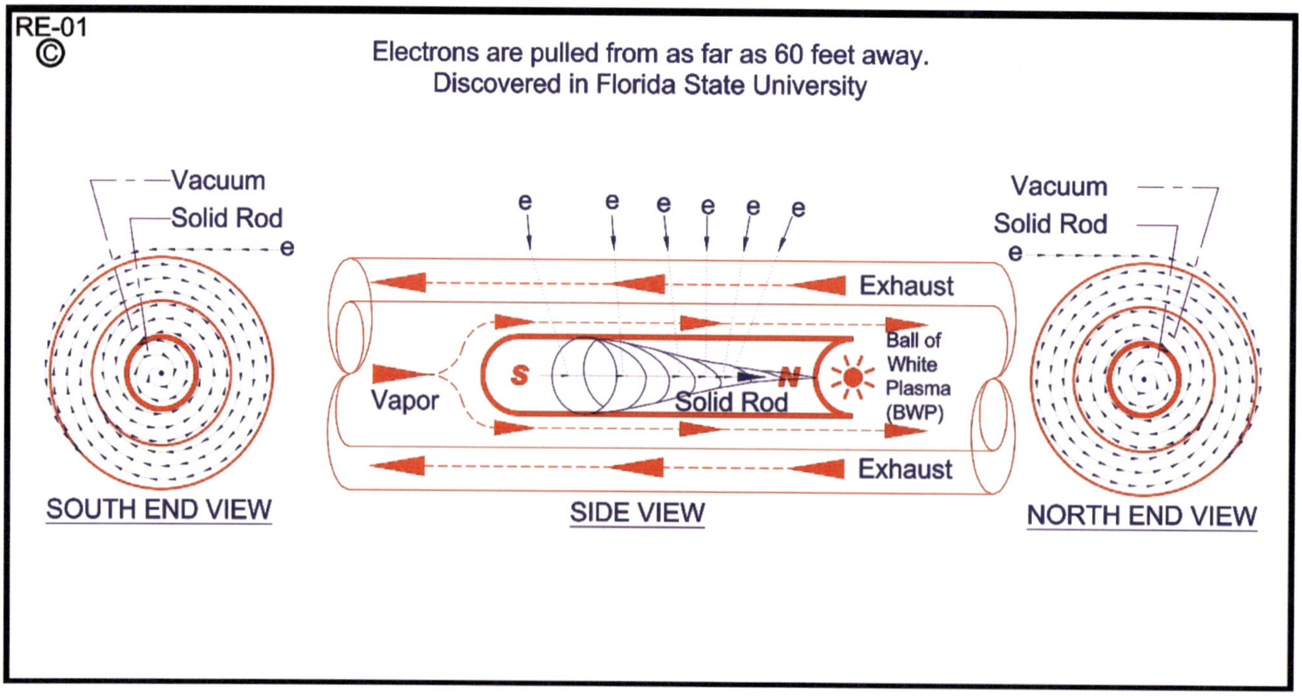

The Transmutation Of Elements Is Real

Transmutation of elements is the restructuring and separation of atomic mass. A particle of sulfur entering the chamber prior to hitting the acceleration rod has an atomic mass weight of 32. As it travels the length of the rod, providing the rod is the right length, it should break down to 32 particles of a substance like or close to hydrogen.

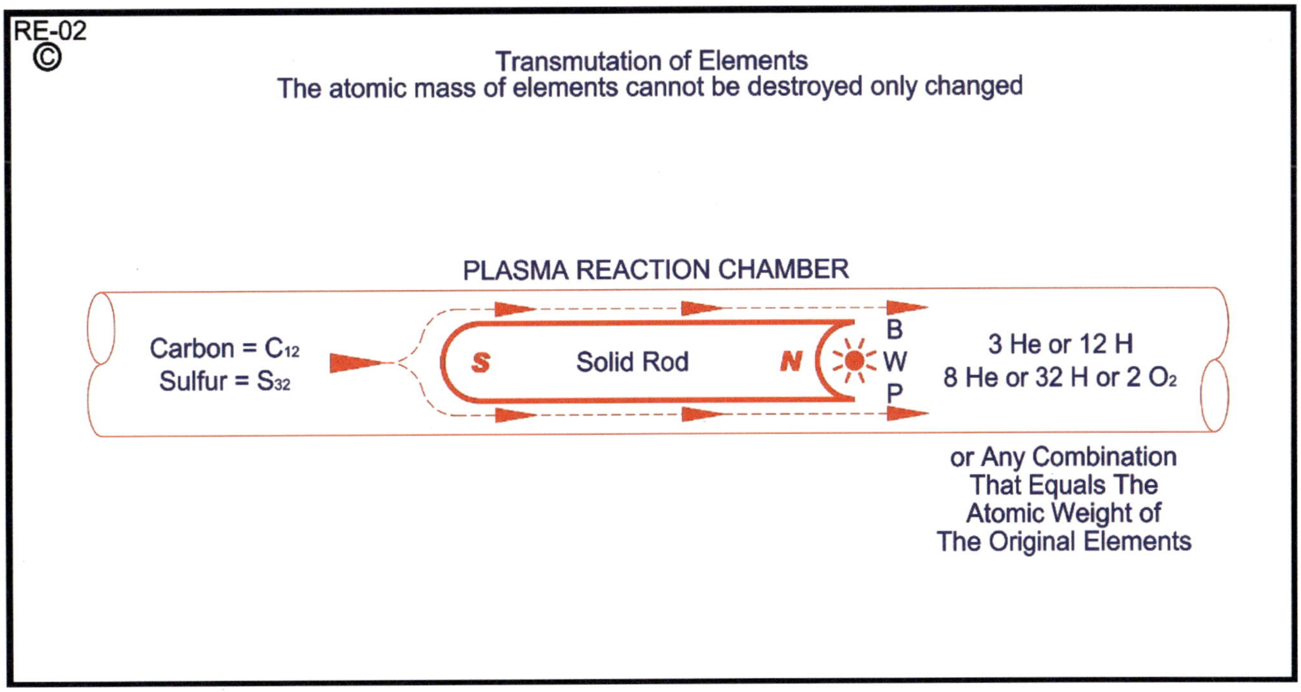

The splitting or transmutation of the elements is caused when the fuel vapor and fresh air reaches the center of the reaction rod. It will split into half of its atomic weight. From the initial halfway point to the end of the rod, it will split the element in half again. This repeats to the end of the rod, delivering a clean light fuel to run your engine or furnace.

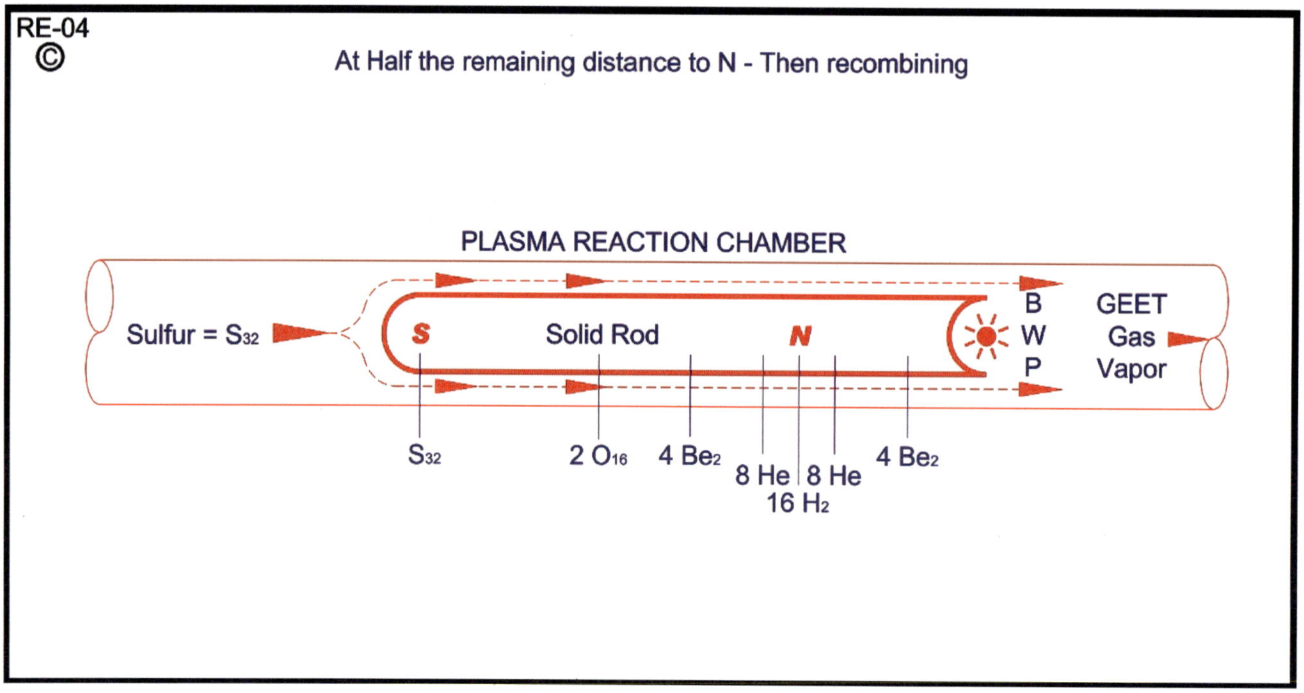

Should your rod be too long, you will have a recombining of elements. This recombining of elements happens after the vapor has achieved a level of maximum separation into its smallest element. It will then become heavier and heavier. During one test, we used a 12-inch rod when we needed a 4-inch rod. The results were a thick black smoke. The black exhaust became so thick that the fire department came thinking we had a tire fire.

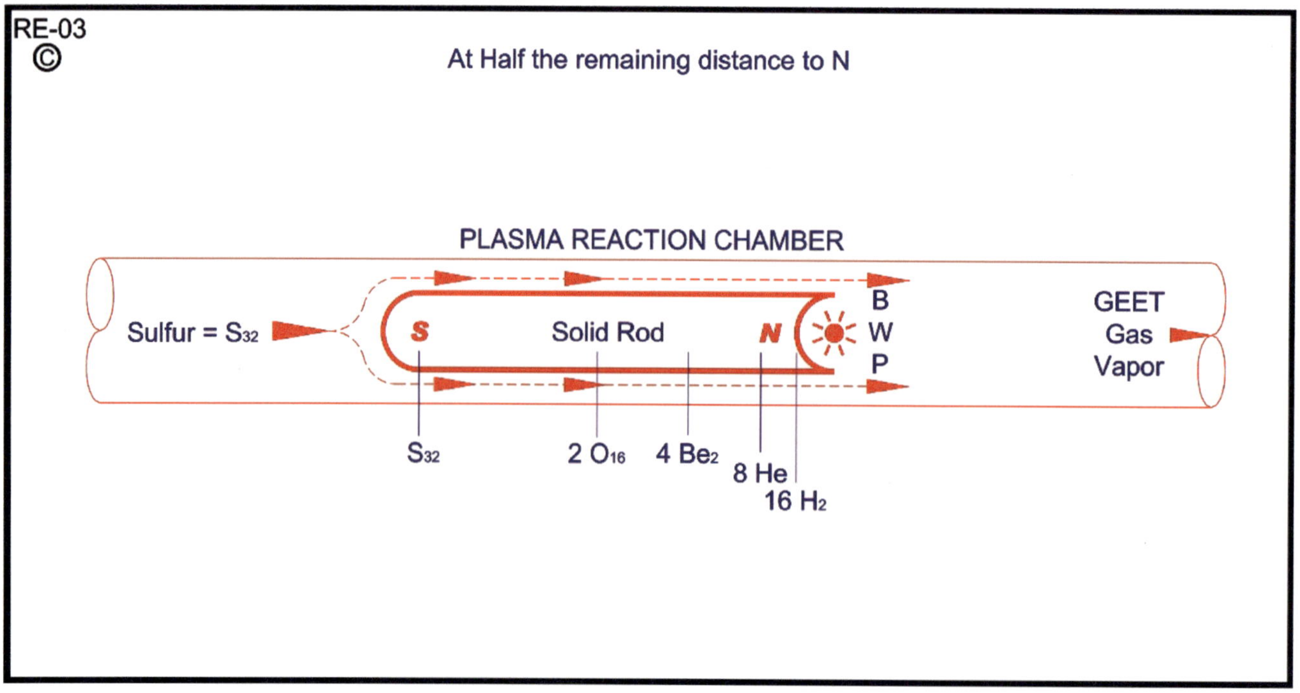

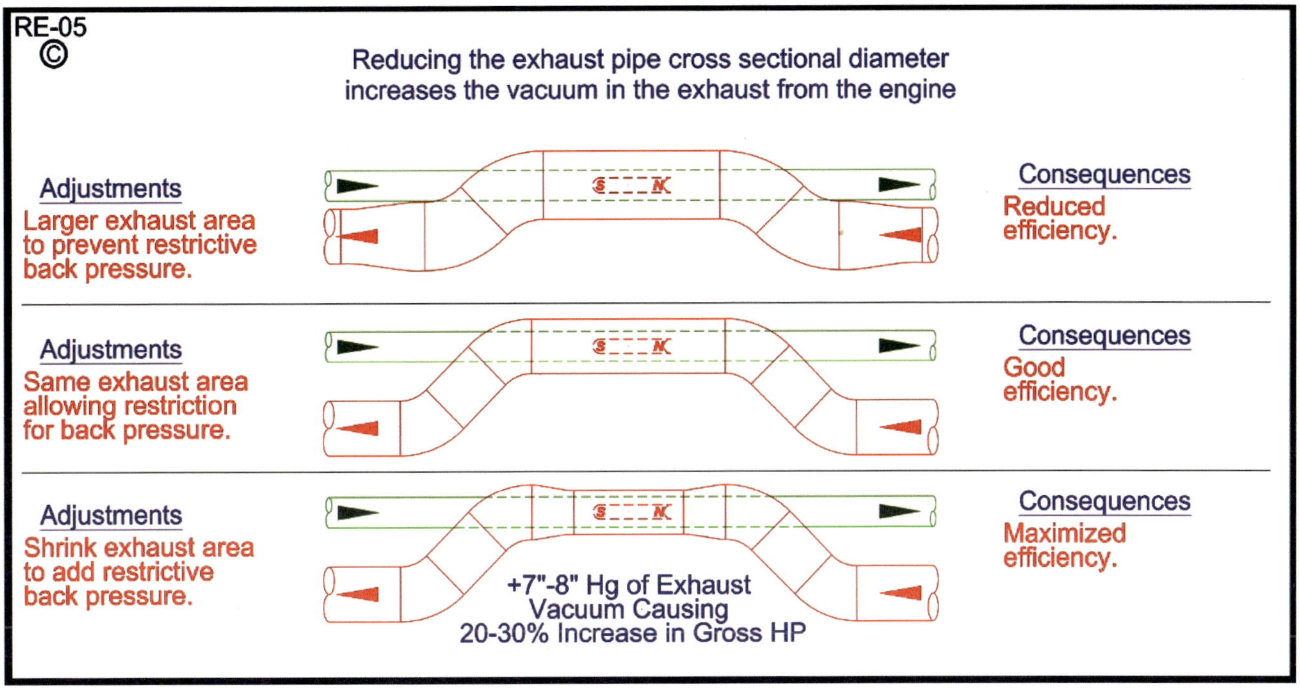

During one of our classes around 2000, some German scientists working with the University of Berlin wanted a private class. We had some problems in communication as I do not speak German and they were not very good with English. I gave them all the information I could and hands-on experience in my shop. After running a GEET retrofitted engine, they were convinced they could go home and build one without any problems.

A month later, I received a call and whoever was speaking at the other end was very excited. They had difficulty relaying to me what they were trying to say in broken English. I asked them to email me a sketch of what it was they were trying to do. He had drawn a picture like that of RE 05, using an arrow pointing at the reactor rod with a question mark. I thought he wanted to know the clearance between the reactor rod and the reactor chamber. I responded for the fuel he was using, it should be about one-eighth of an inch clearance all around.

A few days later, they called to tell me the Volkswagen they were testing at the university had about a 30 percent increase in horsepower. Restricting the exhaust pipe around the reactor was causing a vacuum within the exhaust manifold of the engine. Using basic automotive knowledge, one would think this would have caused a back pressure on the engine, or stopped the engine.

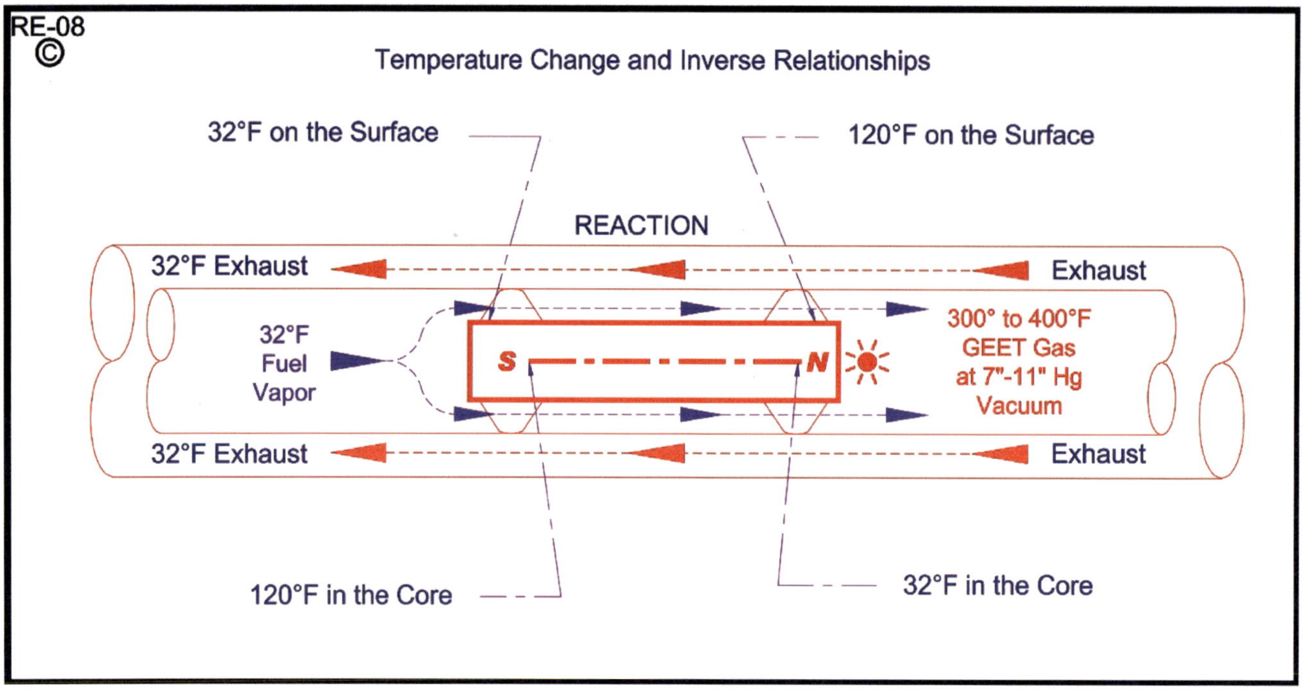

Things were becoming more complex when trying to understand a self-generated plasma unit. During some of our original testing with using a bubbler to create a vapor to feed into the reactor, we had difficulty because the fuel would freeze. Having no books to look into and not knowing where to begin, all we could do was study. So simple yet so complex. I tried to explain, with great difficulty, how the gasoline was freezing in the bubbler jar when the ambient temperature was over 110°F.

The vapor delivered to the reactor rod was usually at or below the freezing point of water. In the beginning, we did not realize the vacuum vapor was spinning at a very high speed. We also did not realize the rod itself was spinning when it could. One day, being in a hurry to do a test, I thought it was only logical if the air was moving into the chamber and hit an obstruction that it might divide and move around the obstruction before coming out the other end. Therefore, I would not put on the three little tabs we had welded at each end of the rods. I was very surprised by the reduction in pollution by allowing the rod to be free moving within the chamber.

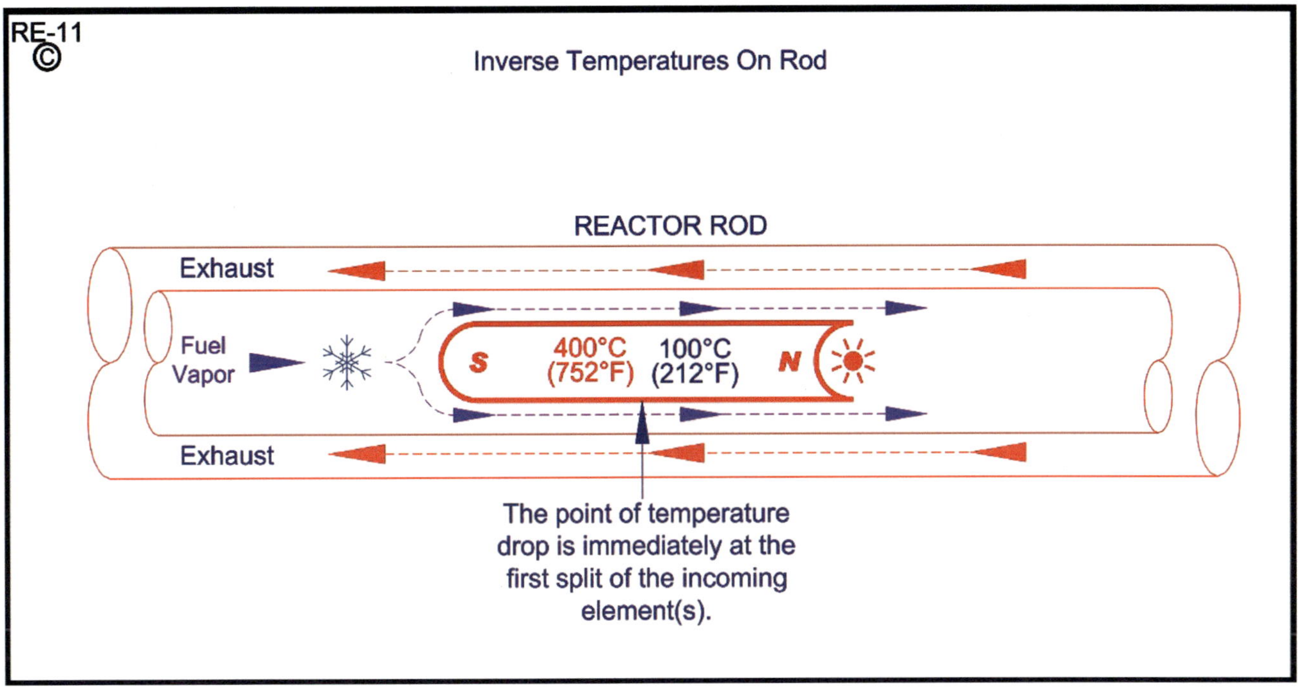

How can you tell the temperature of the GEET rod? One of our students from France used GEET technology to get his engineering degree. During his research, he tested the rod for a hardness based on the Rockwell scale, and determined the temperature had increased over 300°C at the center of the rod. This atomic reaction of tremendous heat pulled the heat from the exhaust and all the metal around it into the reaction. Several rods have been observed with rings, which indicated heat showing where the splitting of the atomic particles was occurring.

The Pull Of Heavenly Bodies

AS-10
©

Orientation And The Pull Of Heavenly Bodies

Baseline = **8 oz fuel**

Factory engine and standard carburetor ran 6 min. 20 sec.

Factory engine and horizontal, North oriented GEET reactor system ran 21 min. 23 sec.

Factory engine and horizontal, North oriented GEET reactor system at Full Moon ran 23 min. 53 sec.

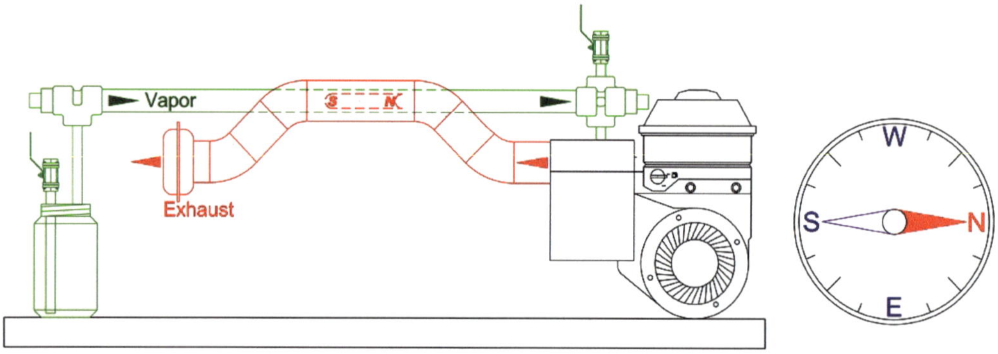

What exactly does a lunar cycle do? When I was young, I used to hear adults talk about the full moon and how strange things would happen. When I got older, the sciences were saying this was a bunch of bunk. Yet for many years in the 1800s, farmers would put paste on small strips of fabric, weigh them down to the table in east and west direction, and glue seeds onto the tape. They aligned the pointed end of the seed to the north, and left it in that position for one full lunar cycle. They claimed the seeds would grow much faster when they were planted in the spring.

I have not tried taping or gluing seeds to a piece of tape, but I had a roll of these seeds given to me back in 1995. When they were planted, the results were quite phenomenal. Could this be true? It is hard to tell if the general public acts crazier on a full moon than a new moon, but I have overheard nurses, doctors, and even policemen complain when they have to work during a full moon. Is this just part of another old wives' tale or is it fact? I have followed many old wives' tales, only to find out they are true. Had this not happened to me personally, I would have found it hard to believe.

A GEET modified generator we delivered was taken apart by the purchaser and was returned because it "didn't run right." After close examination of the engine assembly, it was obvious the customer had taken apart the GEET reactor to see what made it work. Upon closer examination, we realized the rod had been put in backwards as all the seals had been broken. Since we kept records of all of the engines we shipped, I was able to pull up the original records. On this particular engine, we had tested it three times on eight ounces of fuel on each test to make sure it was running properly before we shipped it out.

I replicated the same tests using eight ounces of fuel, and all three times it ran two and a half minutes longer than it did originally. This did not make any sense to me at all at the time. I tracked down the guy who had done the original test and asked him if he could have made a mistake. He said, "No, I always used a stopwatch." Due to the discrepancy in run times, I decided to send the customer a brand new generator, so I could continue testing to see why we had a two and a half minute difference in run times.

To make sure no one shipped this engine out until I was satisfied, I bolted it to the main table where we did our testing. Then I set up a baseline for testing to be done every day at noon, running the same tests to see if there was any difference. We monitored the air temperature and documented the humidity level for each day as we began our testing. After 30 days, we finally saw the pattern only when we typed it into the computer. It became very apparent that a GEET unit will run two minutes and 30 seconds longer on eight ounces of fuel on a full moon compared to a new moon. This is no old wives' tale, but scientific fact.

Maybe before we finish figuring out why, someone out there will get a hold of us and tell us why the GEET units showed these results, but until then, we will continue to share this story the way it happened.

About 20 years ago during a conference in Denver, I drew a picture on the board explaining how we go around the Milky Way and through its belts once every 25,800 years. In my drawing, I show as we go through the galaxy belt the rotation of the earth's magma is the main thing that continues to stay centered, focused, and in line with the center of the Milky Way galaxy. The magma will still spin clockwise, as it always does, but as we go to the south side of the Milky Way, it was then in the opposite direction to north. Since electrons are being pulled to the planet from the sun, as well as a gravitational field caused by the spinning of the planet, the excess energy which is not required to keep the magma's temperature constant is released through the North Pole or South Pole and displayed as the aurora lights.

You will know when poles flip, because the aurora lights will come from both the North and South Pole. Once the flip is completed, the aurora lights will only come from the opposite side of the planet where excess electrons hit the atmosphere particles and create these beautiful lights.

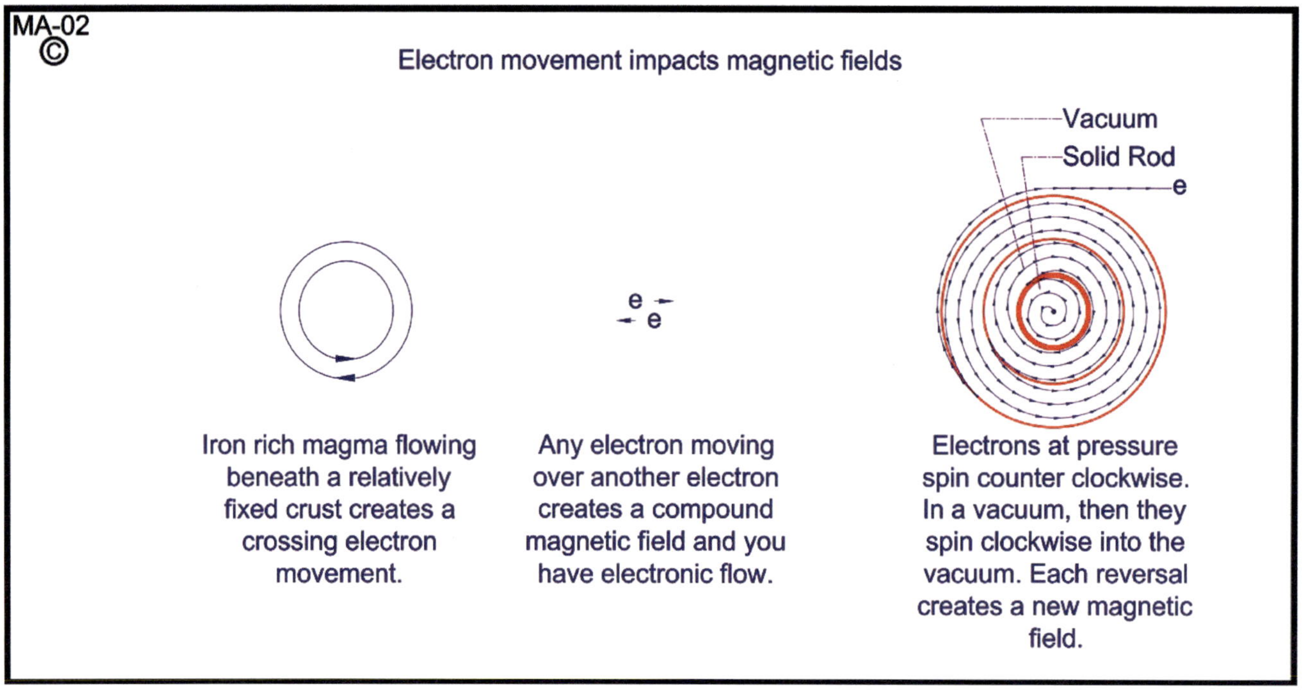

Therefore studying the vortexes within GEET as I have, I must conclude the crust also has a north to south pole at each side of the planet. When the magma is flipped, it will cause the flipping of the two crust sections of magnetic fields. These events will occur with those of plate tectonic movements, volcanoes, and other seismic activities. We know from past archeological evidence man always rebuilds near the ocean. I hope I am wrong, for these events could be catastrophic to low living habitations. It is said history repeats itself, and the remnant evidence is worldwide.

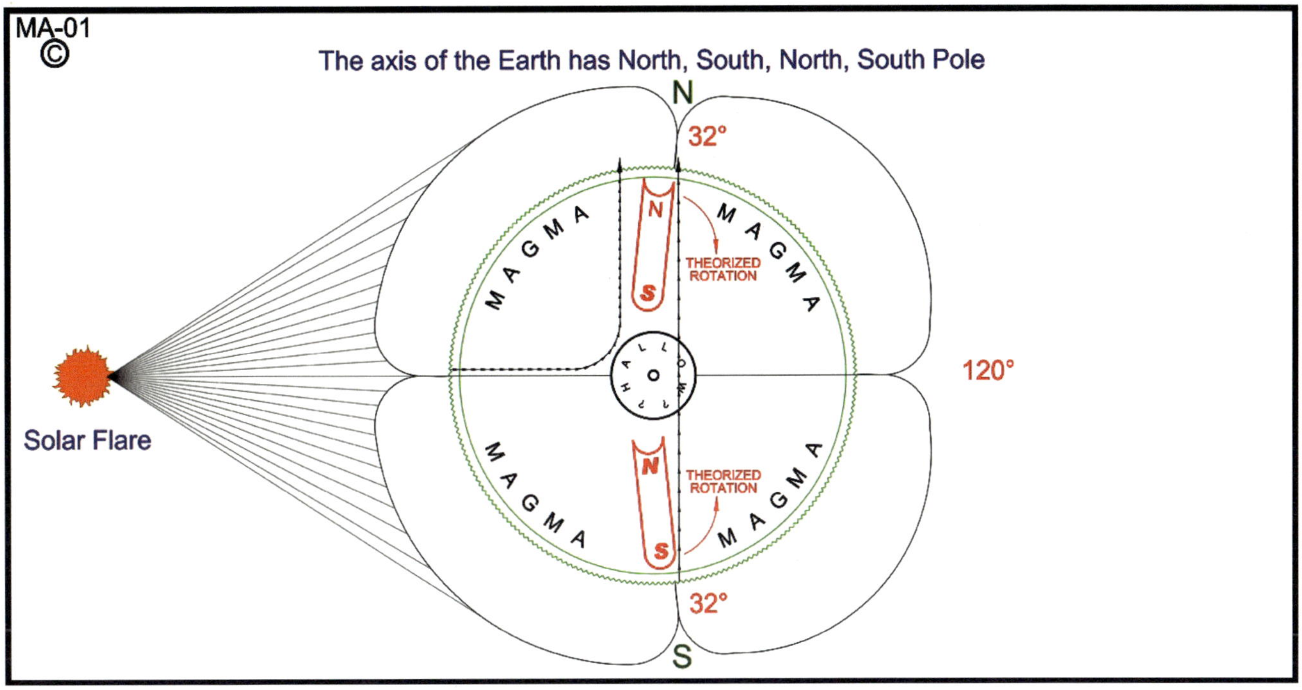
MA-01
©
The axis of the Earth has North, South, North, South Pole
N
32°
MAGMA
MAGMA
N
S
THEORIZED ROTATION
HALLOW
120°
Solar Flare
MAGMA
MAGMA
N
S
THEORIZED ROTATION
32°
S

Let's Make Electricity

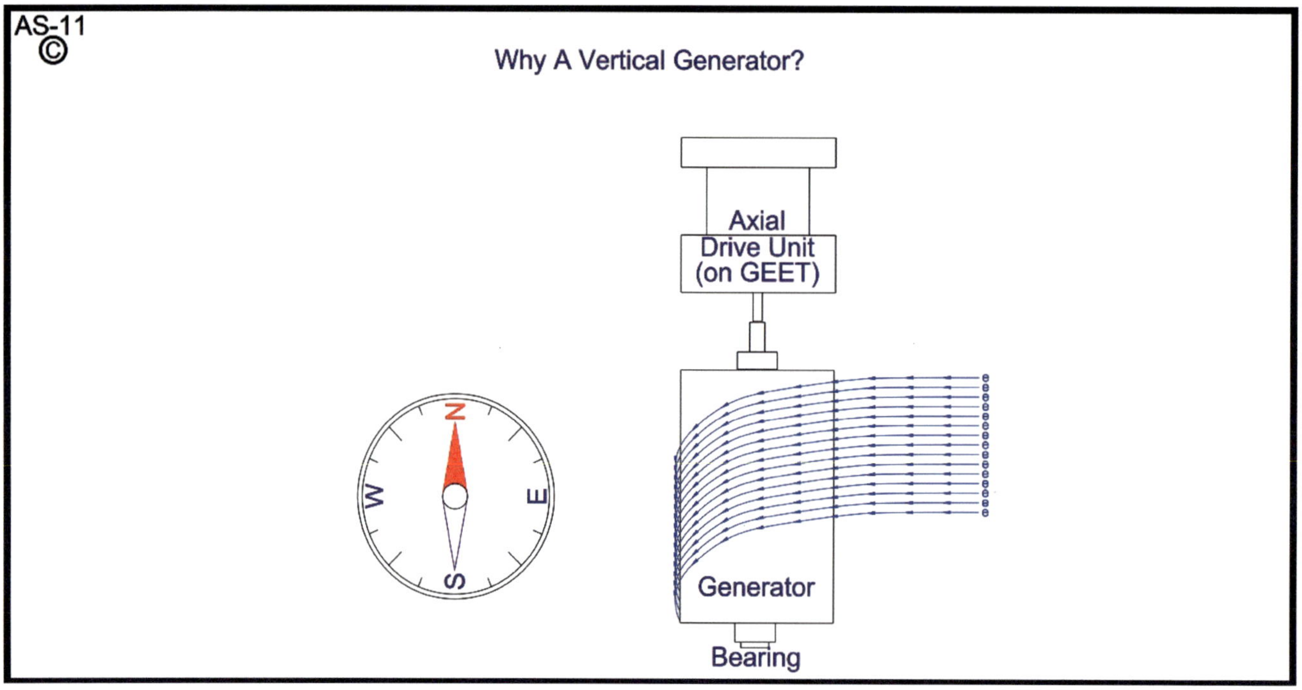

During some of our tests, we purchased generators with high quality bearings to generate electricity using a pulley drive. We mounted the generator in a vertical position instead of horizontal.

MI-09
©

Electrical Arcs Travel in Straight Lines In a Vacuum

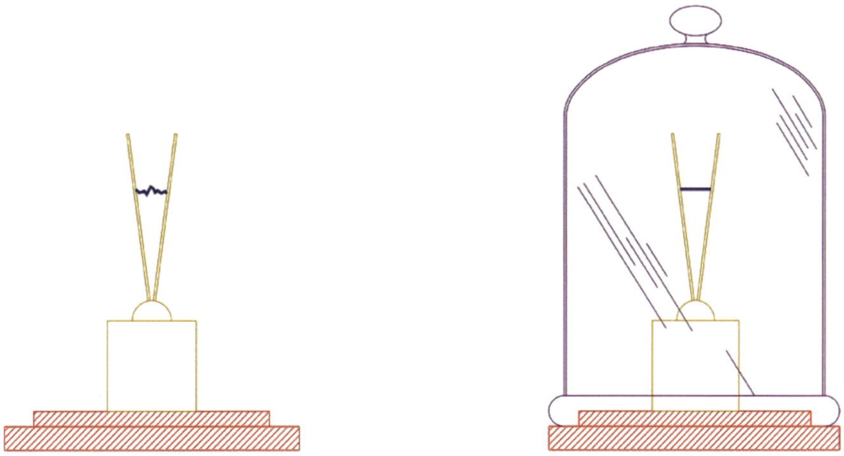

Jacobs Ladder in Open Air

Jacobs Ladder in Vacuum in Bell Jar

I think you will find the results quite entertaining. First, run the engine in normal ambient conditions. Then compare it to what you get when it is in an enclosure of half-inch Plexiglas with about five inches of vacuum. In a vacuum, electricity moves in a straight line and does not require the same amount of energy as it does in atmospheric conditions.

The theory is that white holes are found at the other end of a black hole. Energy is pulled into its self and matter is ripped apart to its lowest possible atomic mass weight. Then at the other end of the black hole, it falls into a void and becomes a ball of white plasma light. When looking at this with the naked eye, it appears to have a blue sheen around the ball of white light.

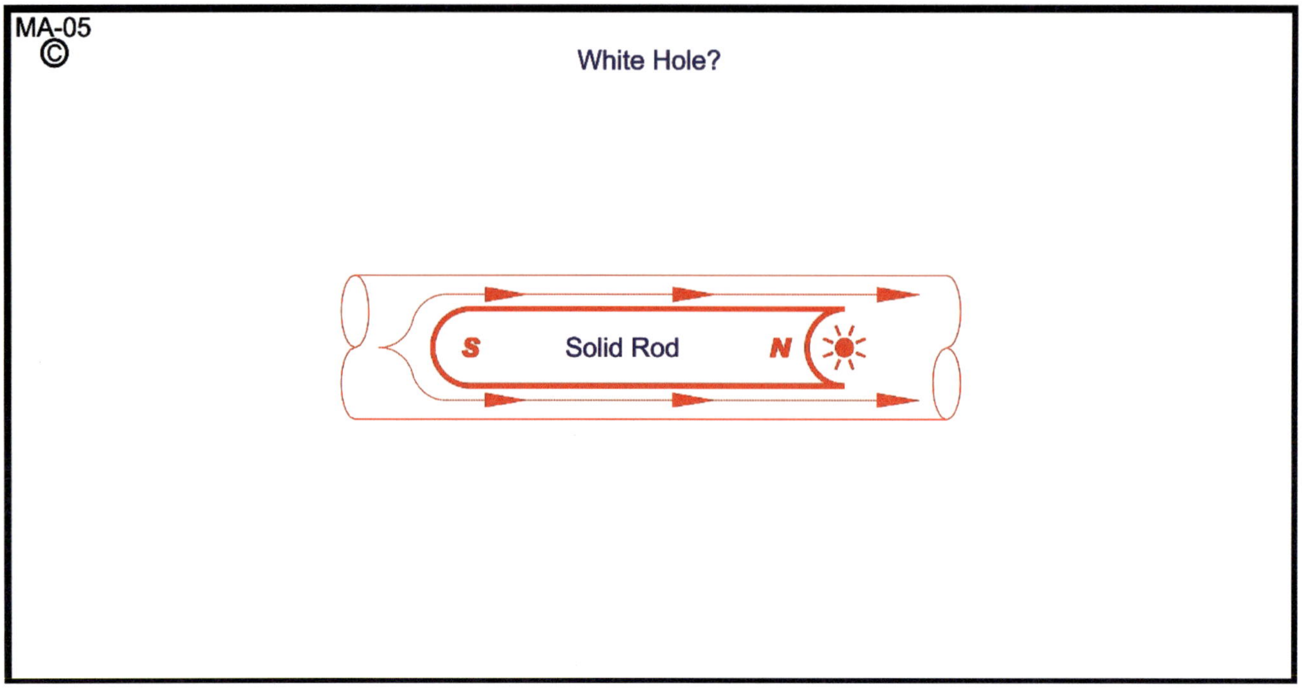

This blue sheen, much like an auric effect, indicates a healthy field of energy. While running tests using a glass chamber, even high sulfur crude oil had been transmuted to a new fuel which came out of the end of the reactor rod as a ball of white light. On one such test at a very high speed, the entire exhaust pipe became transparent. It looked like a scene out of the movie "The Philadelphia Experiment." Human nature caused me to have an immediate reaction, wanting to put my hand where the rod and exhaust pipe had been or might still be. Fear told me not to try to pass my hand through the field.

There is much to learn about white holes and how they apply to energy. There is also much to learn about black holes, but this will give us a place to start. Before we go too far afield, we must learn what it is we are playing with.

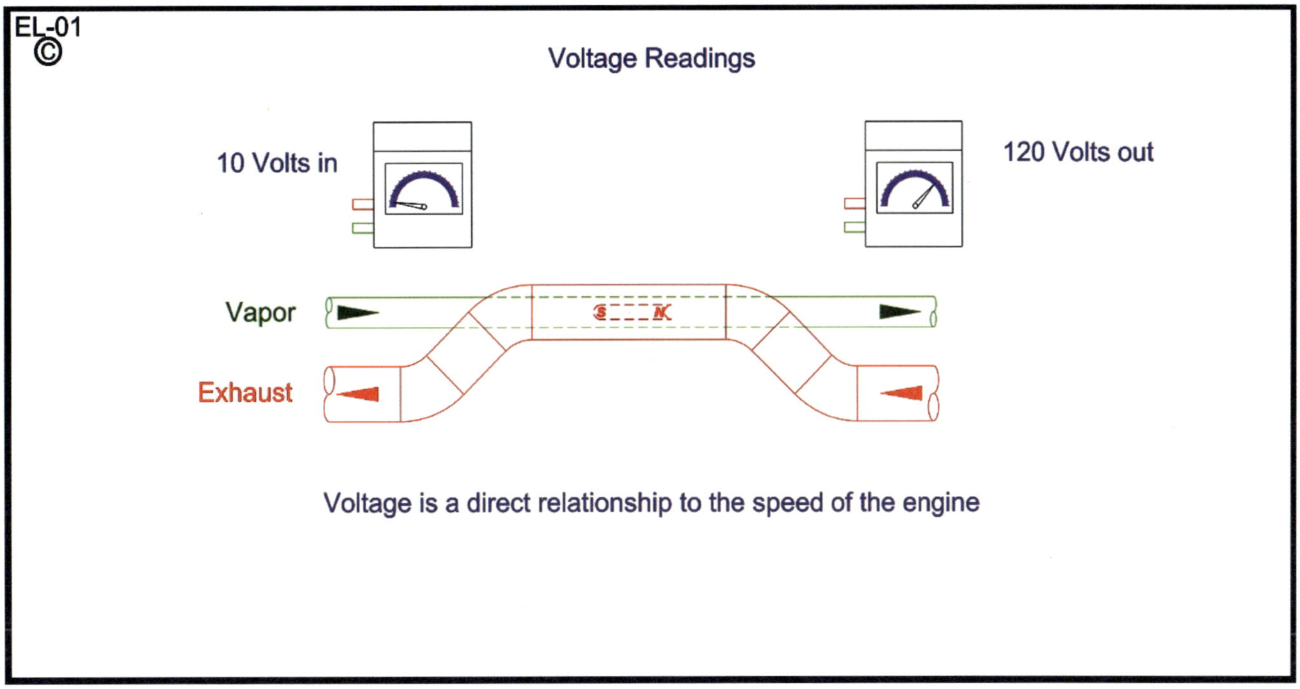
EL-01
©
Voltage Readings
10 Volts in
120 Volts out
Vapor
Exhaust
Voltage is a direct relationship to the speed of the engine

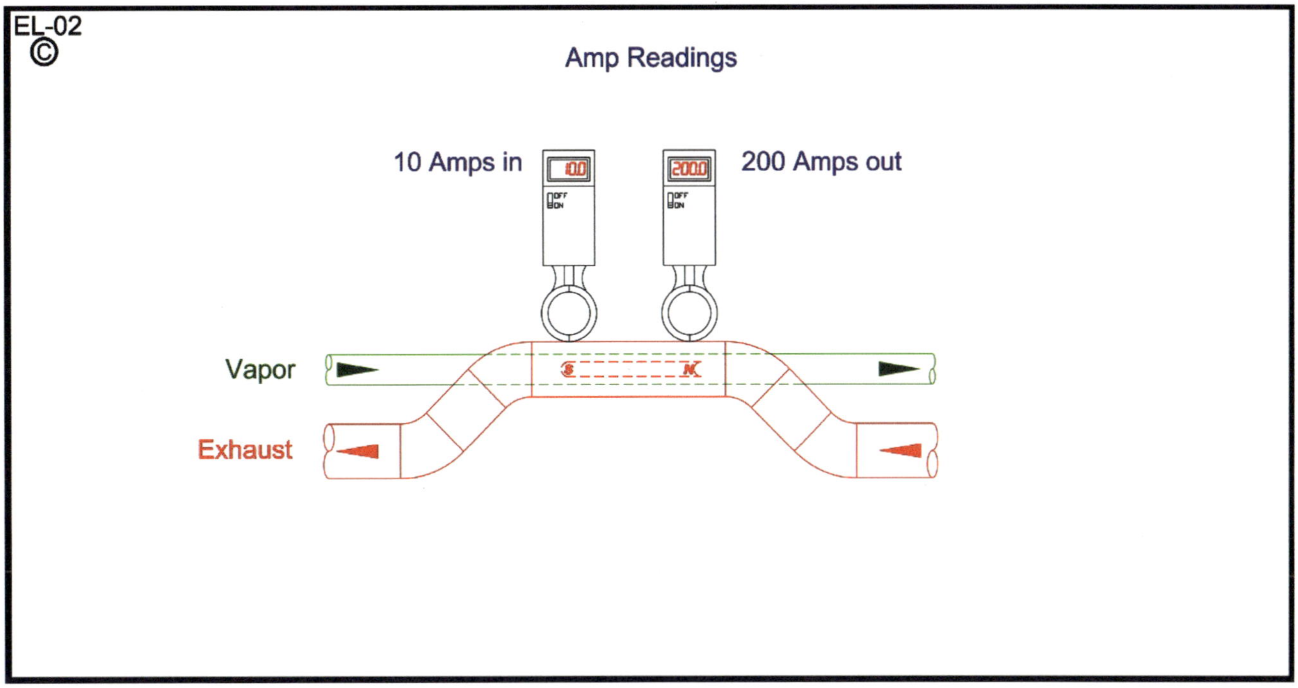
EL-02
©
Amp Readings
10 Amps in
200 Amps out
Vapor
Exhaust

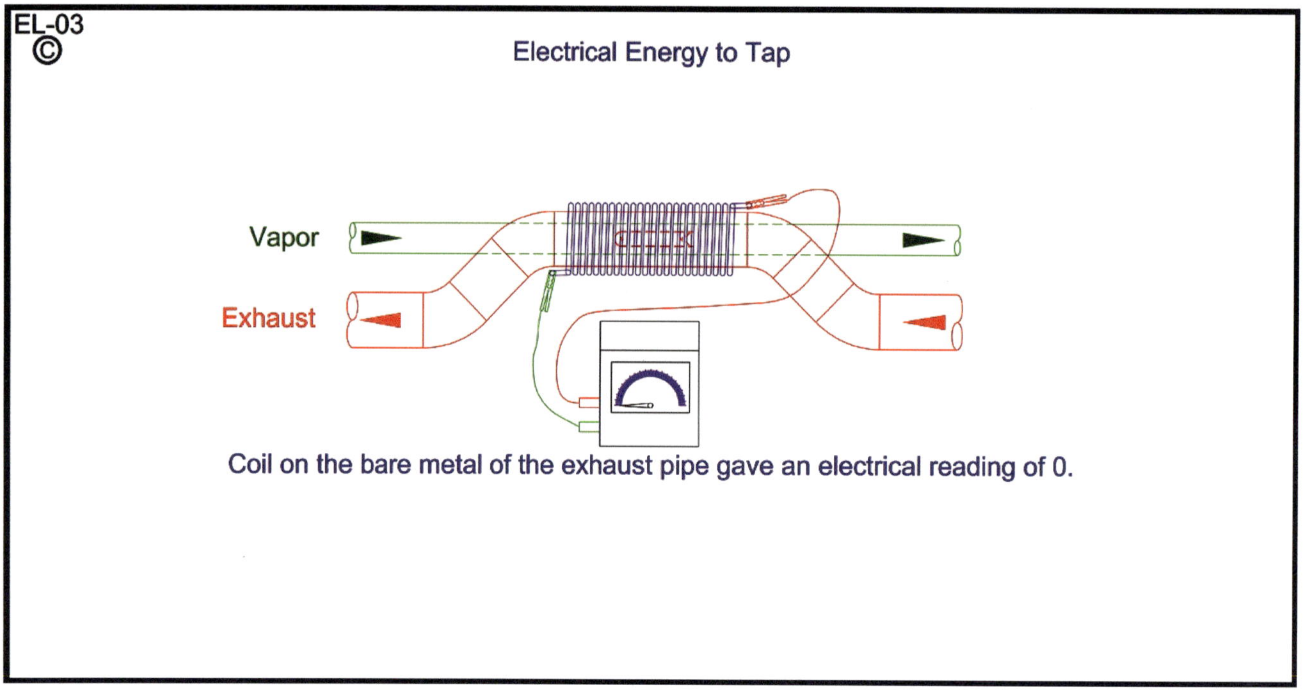

By now, some of you must be questioning why I have called the unidentified fields of energy electromagnetic gravitational fields? The answer is quite simple. We have seen around the shaft area of the exhaust, displays of magnetic fields of energy as well as displays of AC and DC current.

Some people have reported over 200 Amps of power in the air around the north end of the rod. I did say "around" the exhaust area. This is not to be confused with the meter clamped around the unit, but beside it.

Levitation?

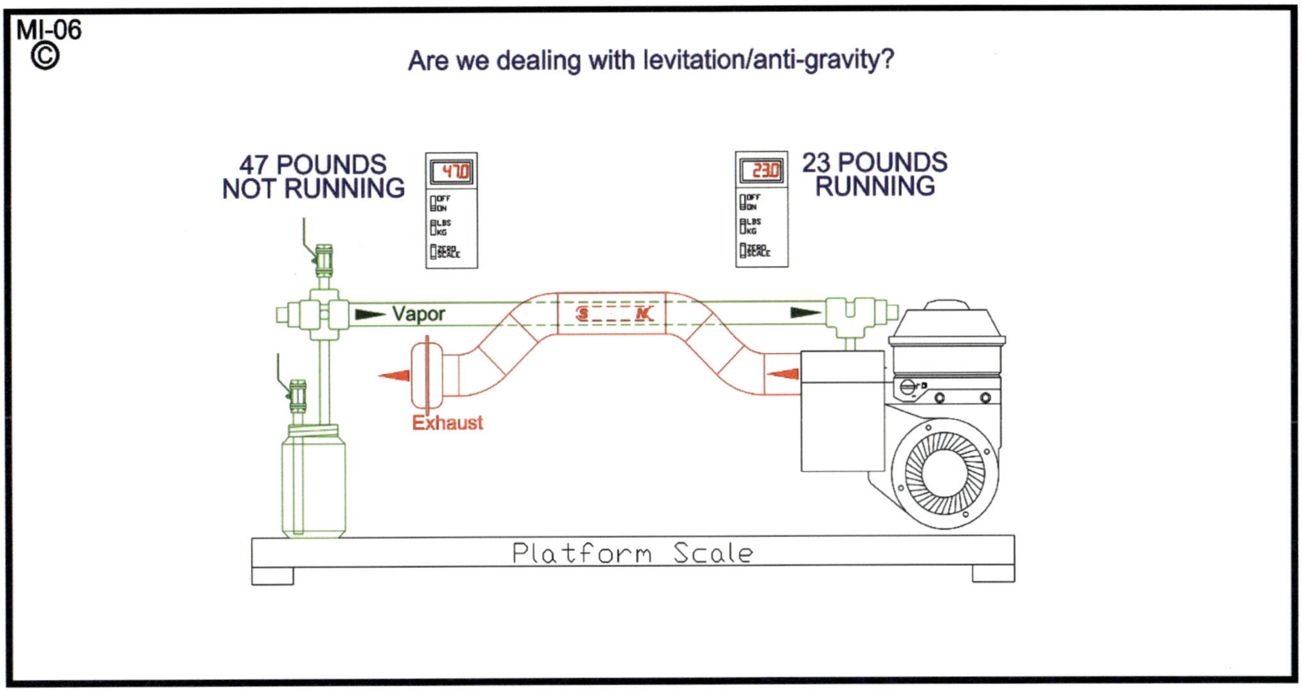

Why do I say electromagnetic gravitational fields? Back in 1993, I tested a 47-pound engine with a GEET unit while it sat upon a scale, and the engine only weighed 23 pounds when I revved it up. This might explain why it takes a lot of effort to keep the engine on the ground and in one spot when demonstrating it.

When I was doing testing in 1994, I had bolted a 10 hp engine onto a 140 pound shop table. When the engine started, I needed to use the restroom. When I returned, the engine was still running and down the driveway. It took four of us to move the table-mounted engine back into the building, where we then secured the table to the concrete floor.

Home Energy Units

As you begin to understand the physics behind why GEET works, you begin to understand many other things are not properly and scientifically broken down for the layman. The same way a tepee causes air conditioning, so does the movement of air in the GEET furnace. It comes down and swirls faster and faster as it gets closer to the bottom, but in a clockwise direction only. When it gets to the bottom, it has to reverse direction and spin upward in the opposite direction while creating an electromagnetic gravitational field. This causes a chilling effect on the bottom of the chimney that builds up as ice.

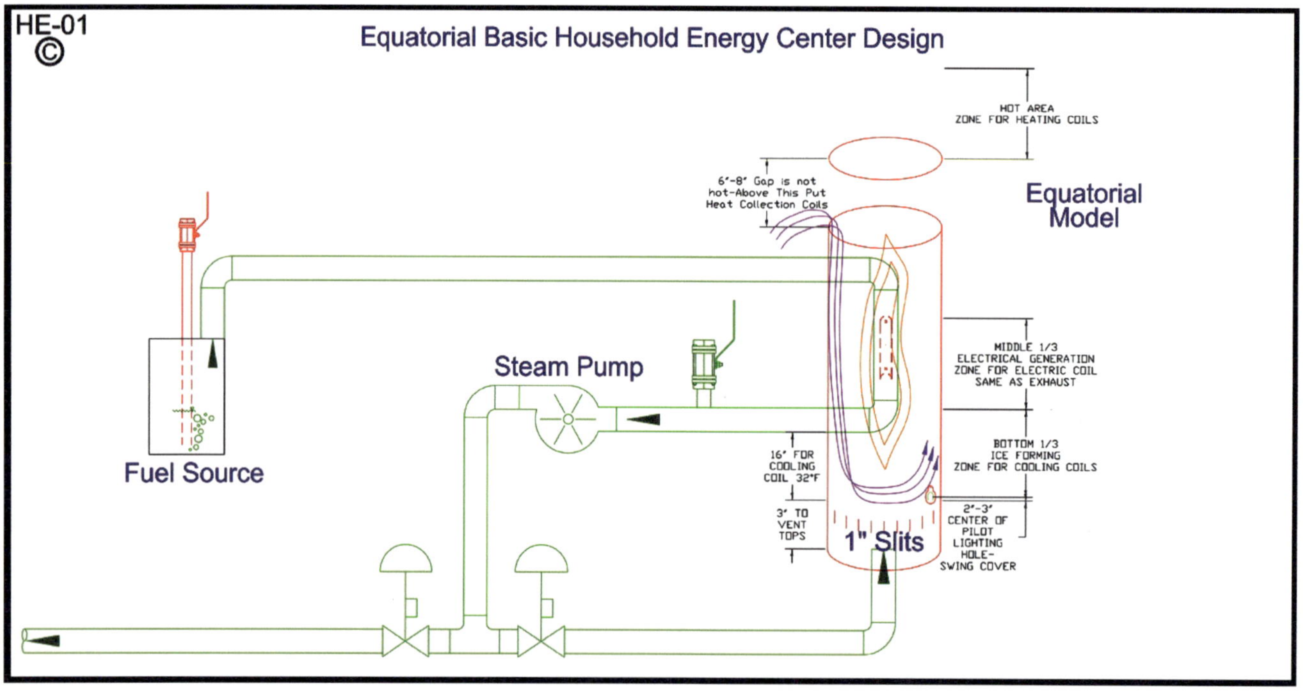

If you allow the ice to build up too much, it stops any air from coming into the lower holes and shuts itself off. Putting a coil around the bottom of the GEET furnace pipe allows liquid or air to move through the coiled pipe to maintain ambient temperature or change it. To lower the pipe, one can maintain the temperature of the heat coming out of the top of the pipe while disposing all the cooling being caused by the lower part of the pipe. This cooling can be anywhere from 0°F to 30°F. Some people have a use or need for this cooling as in food storage or personal cooling.

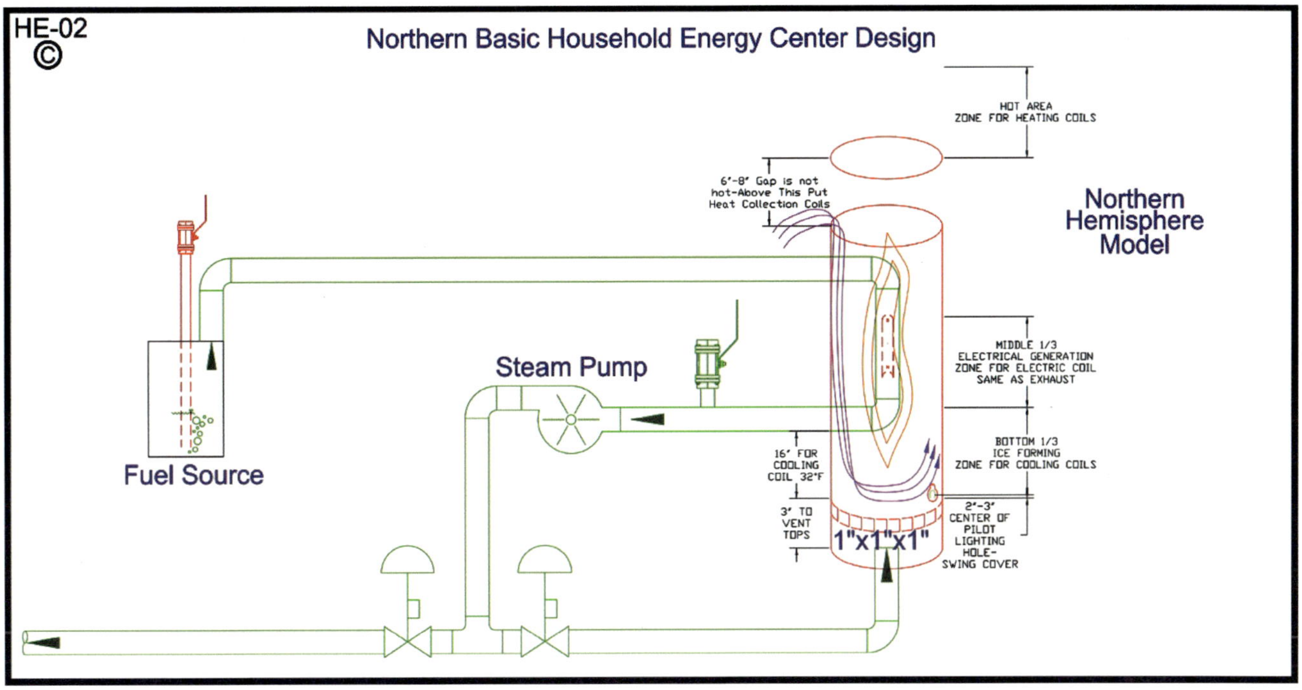

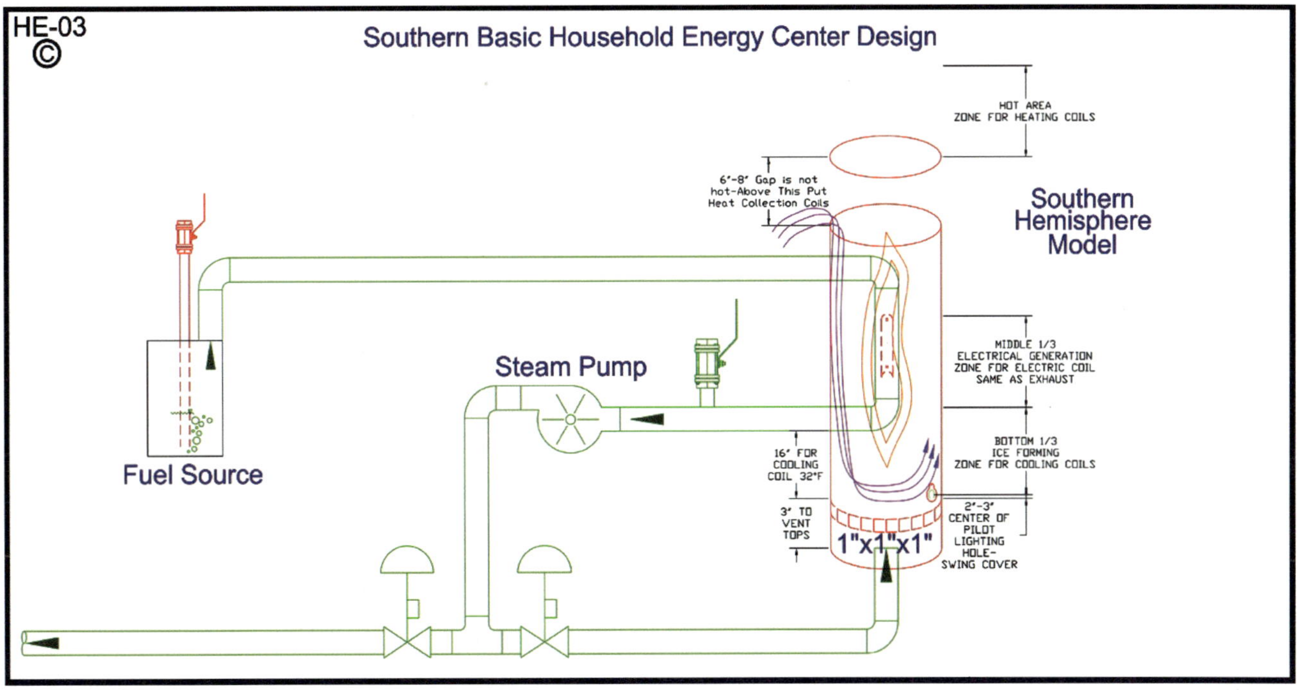

If you happen to live up in the northern part of Alaska or down in Antarctica, you really do not want extra cooling so this could create a problem. The same holds true at the equator where the people want the cooling and they do not want the heat. As they say, you can't please them all.

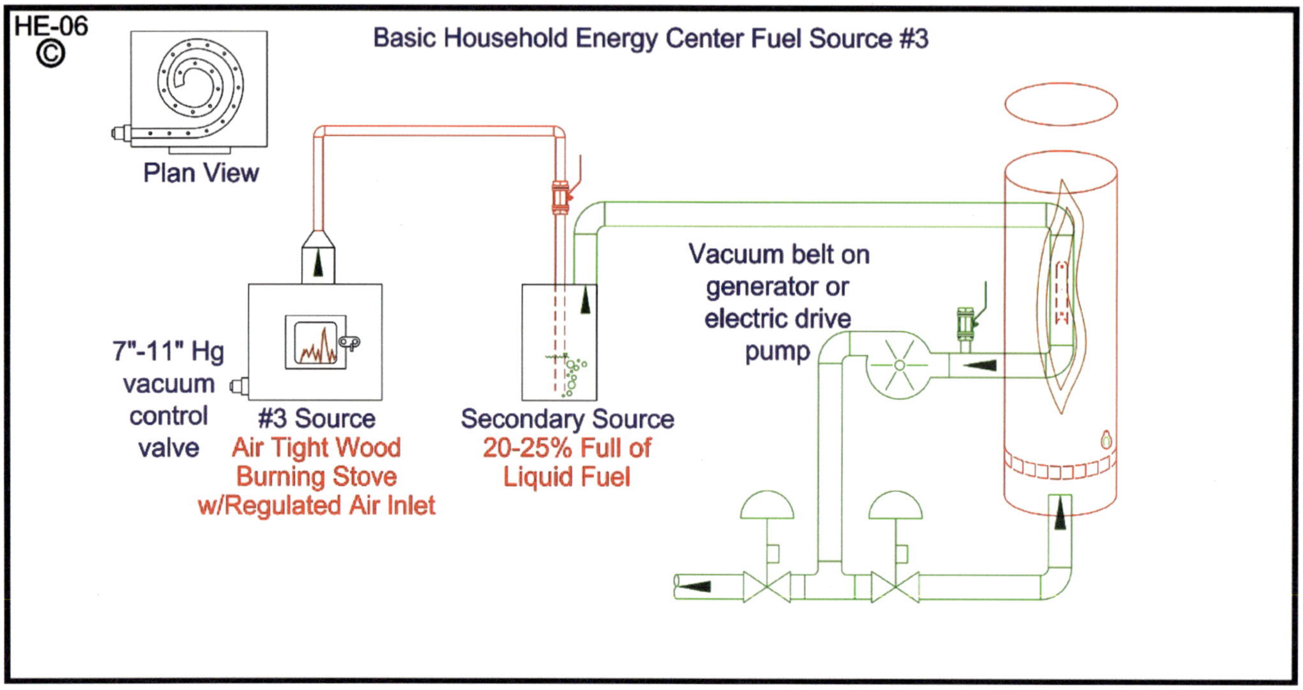

There are several ways to make a GEET. Every model of every piece of equipment on the planet using fuel will be slightly different. Every fuel used in those applications will have different requirements. When heating with tires, coal, or wood, you need an airtight woodstove or chamber that will help to swirl air through the smoke vapor.

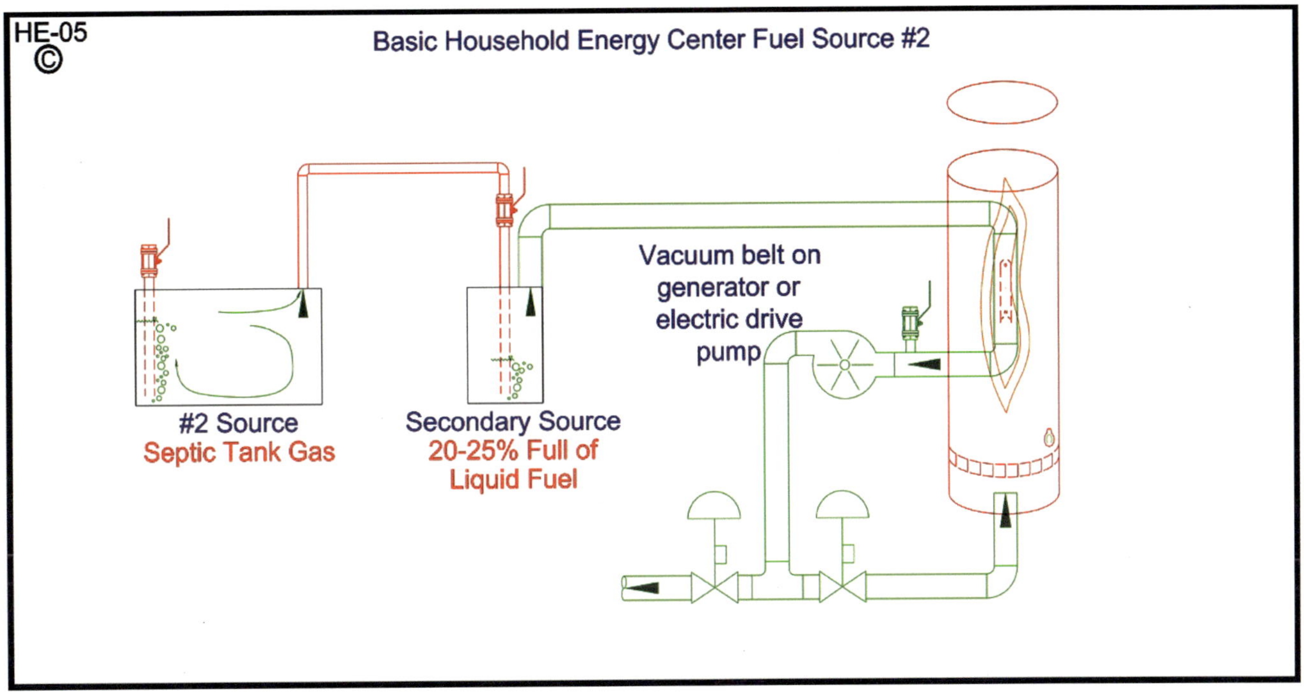
HE-05
©
Basic Household Energy Center Fuel Source #2
Vacuum belt on generator or electric drive pump
#2 Source
Septic Tank Gas
Secondary Source
20-25% Full of Liquid Fuel

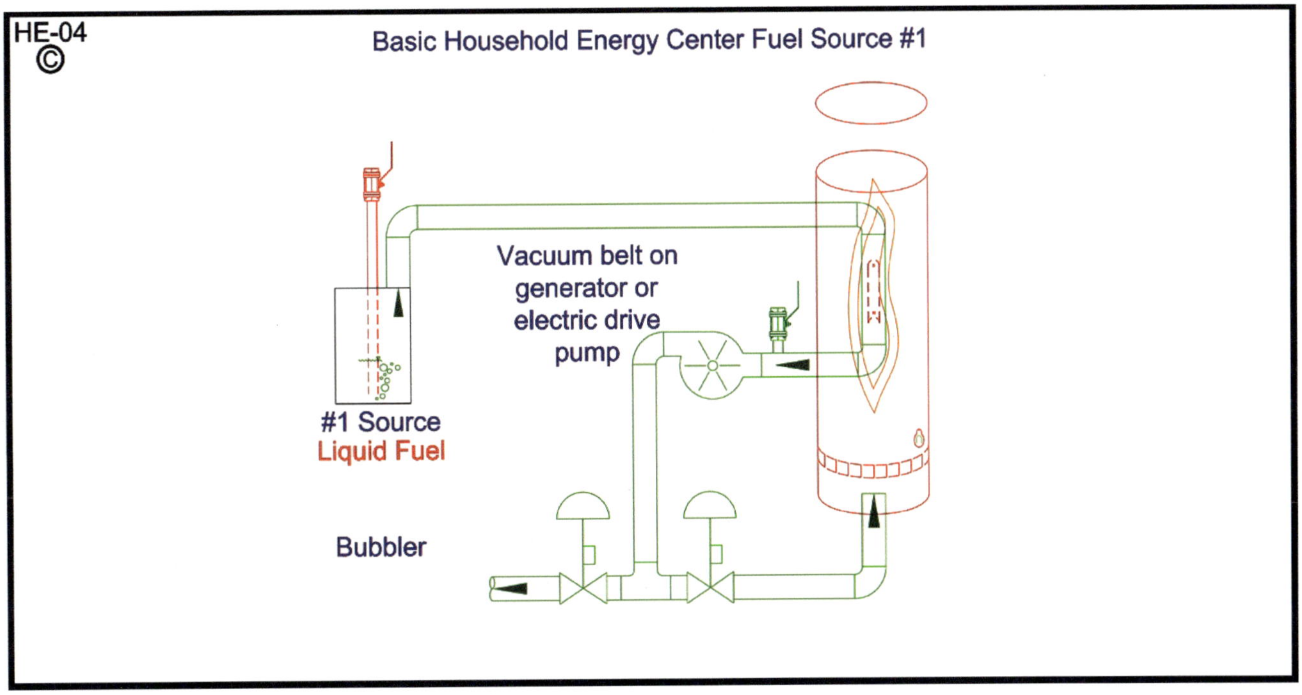
HE-04
©
Basic Household Energy Center Fuel Source #1
Vacuum belt on generator or electric drive pump
#1 Source
Liquid Fuel
Bubbler

Tiny Bubbles Contain Power

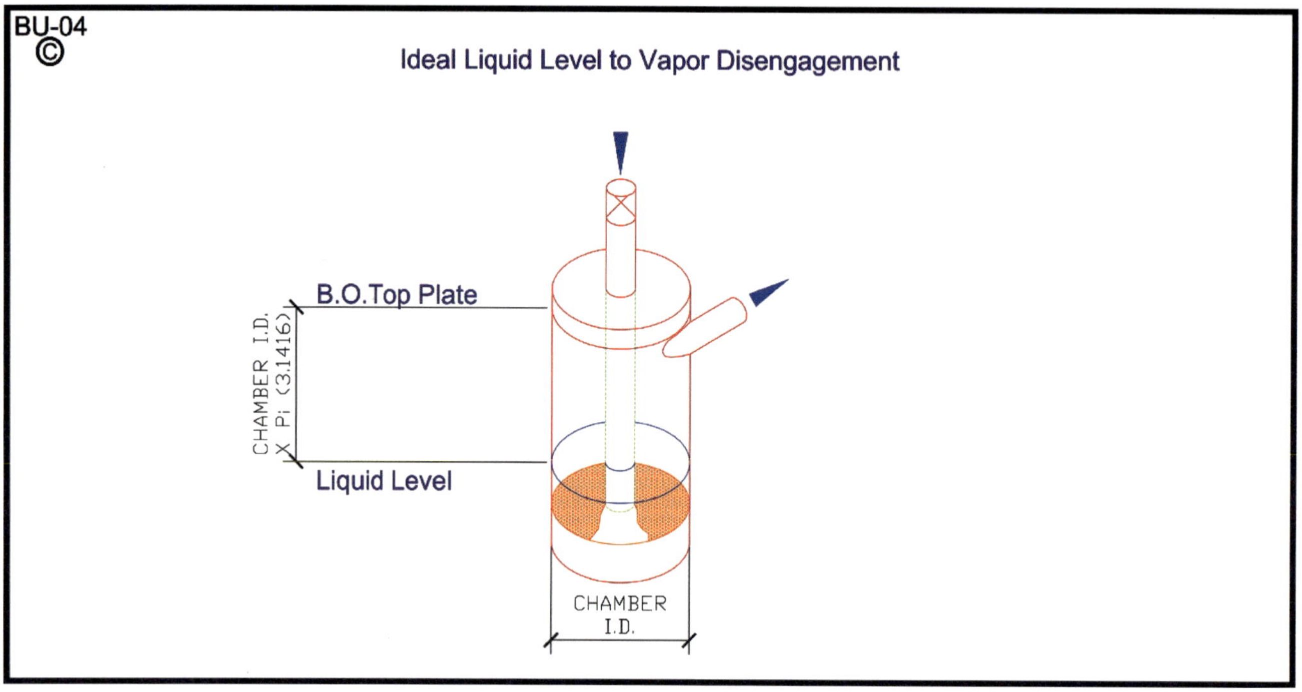

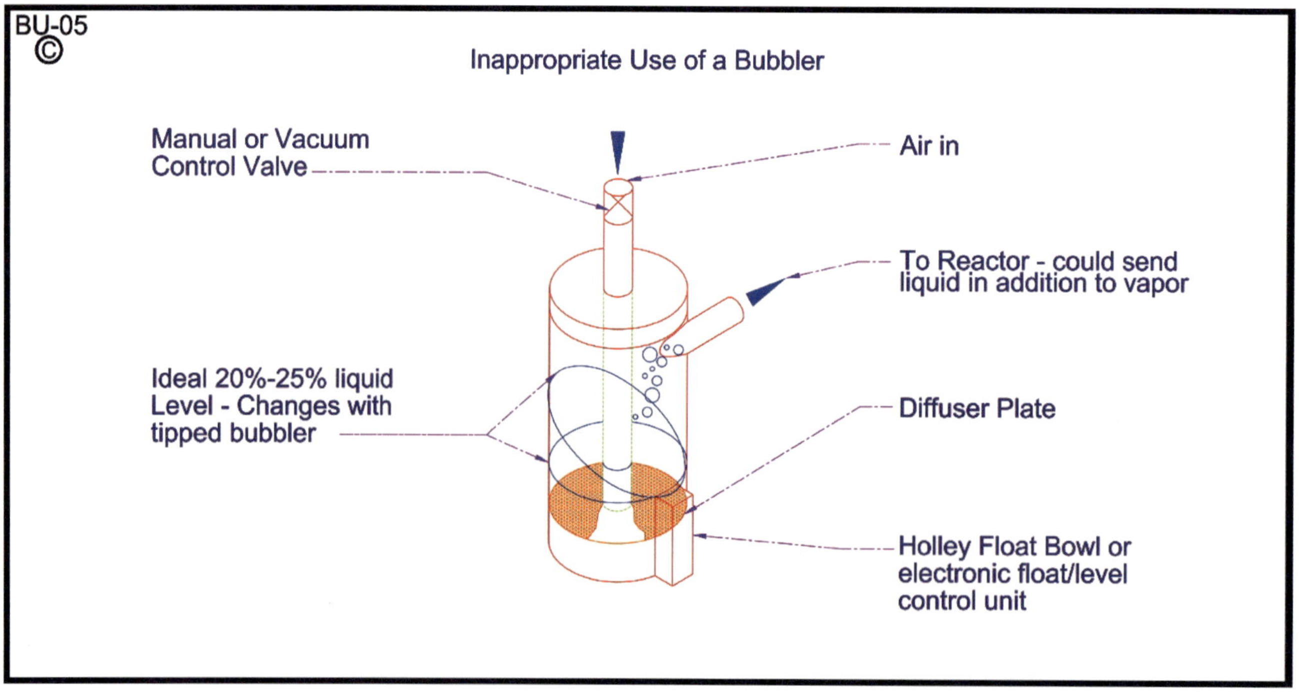

With bubblers, many of the pictures in our old class book were wrong. You should always try to get the air intake as close to the center as possible and ending about one-eighth to one-fourth of an inch above the inside bottom. The closer to the center of the bubbler cylinder, the more evenly dispersed the electron movement will be. I am referring to the movement of air as an electron movement. Many other unidentified parts of electricity are spinning clockwise going downward into the vacuum. They reverse to go clockwise when going up.

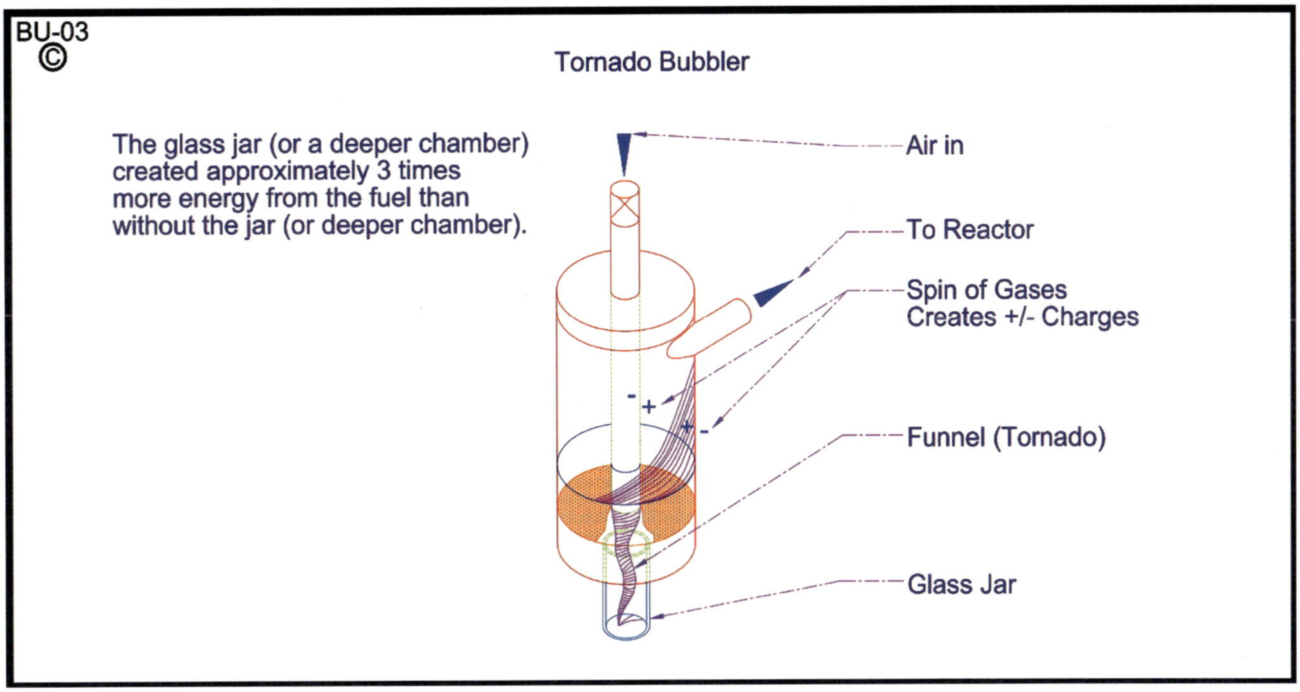

When the vacuum from the top of the chamber pulls this air from the bottom of the chamber upwards, it also spins clockwise in the upward direction towards the vacuum. This is just the opposite of the spin going downward in a clockwise direction causing electromagnetic fields to occur within the vapor. These ionized particles are then fed into the tube going to the reactor inlet.

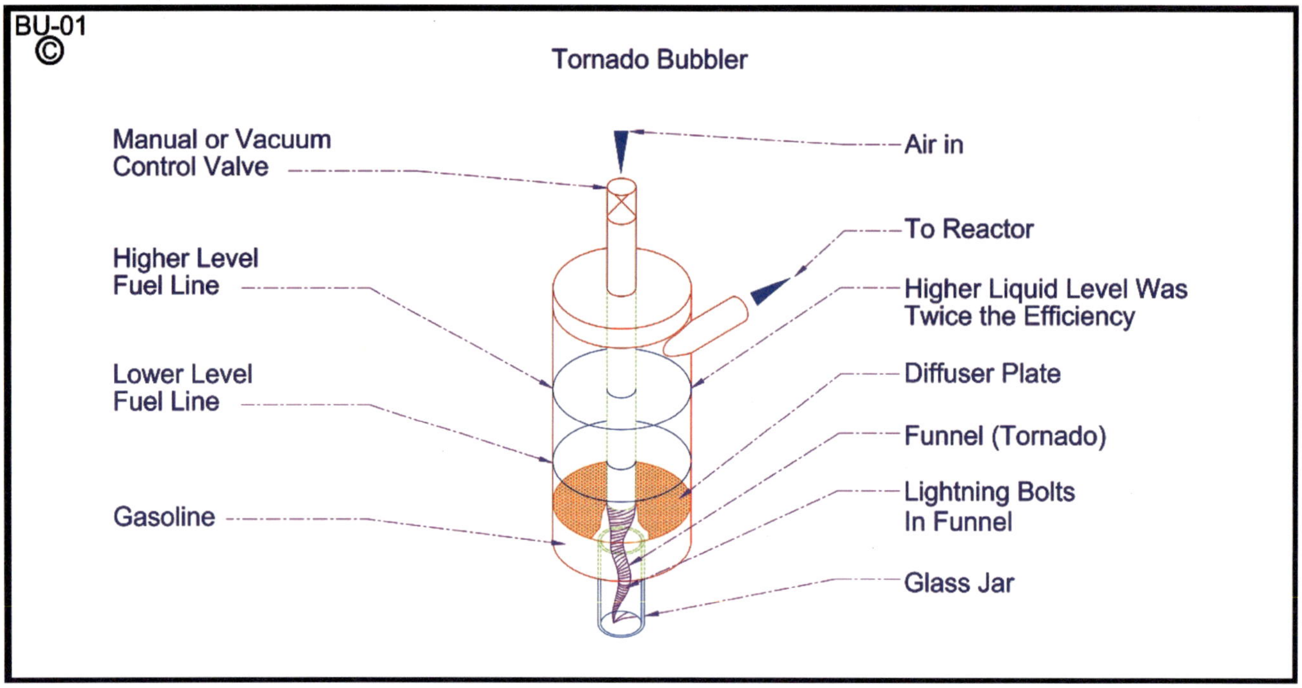

The heavier ends of each particle are being pulled ahead of the lighter ends and causes a particle separation to happen more quickly. One of the quirky discoveries made is when you add sugar or salt to the liquid fuel in the bubbler, it lengthens the molecular chain and increases the power level of the fuel. This provides better mileage and, in most cases, less pollution. This can be rather humorous at times when mechanics see you putting in two or three small bags of sugar into the fuel chamber for the demo engine, since they learn sugar will destroy the engine. They all overreact. Once the engine starts, they all seem to be overjoyed to see their textbooks were wrong.

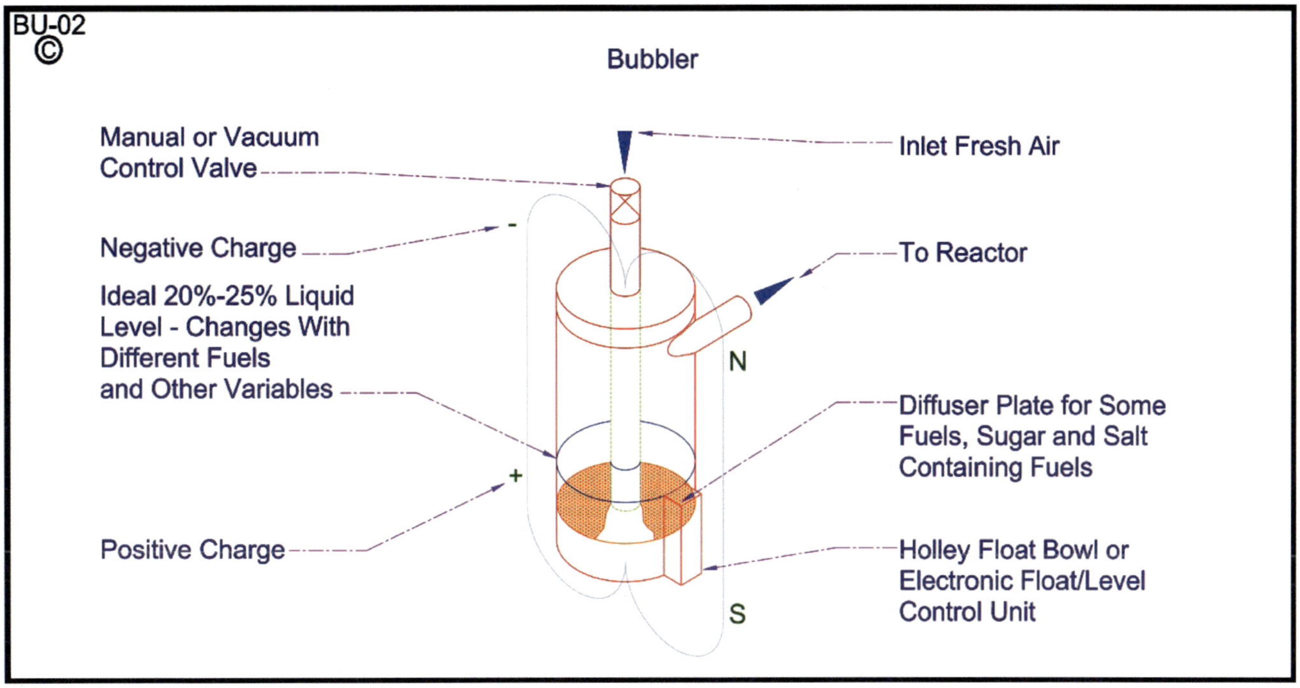
BU-02
©
Bubbler
Manual or Vacuum
Control Valve
Inlet Fresh Air
Negative Charge
-
To Reactor
Ideal 20%-25% Liquid
Level - Changes With
Different Fuels
and Other Variables
N
Diffuser Plate for Some
Fuels, Sugar and Salt
Containing Fuels
+
Positive Charge
Holley Float Bowl or
Electronic Float/Level
Control Unit
S

Made in the USA
Las Vegas, NV
09 November 2026

57738627R00102